HEROIC ANIMALS

Also by Clare Balding

My Animals and Other Family
Walking Home

FOR CHILDREN

The Racehorse Who Wouldn't Gallop
The Racehorse Who Disappeared
The Girl Who Thought She Was a Dog
The Racehorse Who Learned to Dance

HEROIC ANIMALS

100 AMAZING CREATURES GREAT AND SMALL

CLARE BALDING

JOHN MURRAY

First published in Great Britain in 2020 by John Murray (Publishers)
An Hachette UK company

1

CIP catalogue record for this title is available from the British Library

Hardback ISBN 978-1-529-34382-3
Trade Paperback ISBN 978-1-529-34383-0
eBook ISBN 978-1-529-34385-4

Text design by Janette Revill

Typeset in Goudy Old Style 11/13 pt
by Palimpsest Book Production Limited, Falkirk, Stirlingshire

Printed and bound in Great Britain by Clays Ltd, Elcograf S.p.A.

John Murray policy is to use papers that are natural, renewable and recyclable products and made from wood grown in sustainable forests. The logging and manufacturing processes are expected to conform to the environmental regulations of the country of origin.

John Murray (Publishers)
Carmelite House
50 Victoria Embankment
London EC4Y 0DZ

www.johnmurraypress.co.uk

For the everyday heroics of the animals who have been there when we needed them most

In memory of Archie (2005–20)

Contents

Introduction

I have always believed that I have been shaped more by the animals in my life than the people. Through countless acts of love and kindness, or in offering encouragement as I endeavoured to find a way to communicate with them, animals have made me who I am. Whether it was Candy the Boxer, who was my early protector and nanny, Valkyrie the Shetland pony who did her best to teach me manners, Volcano the Welsh Mountain pony who taught me patience, Frank the Heinz 57 pony who was my first real love or Henry the runaway who taught me how to be brave, I owe everything to animals.

I was not quite born in a stable and schooled in a kennel but near enough. I grew up surrounded by dogs, ponies and horses. They were way above me (and my brother) in the pecking order of family importance and I was perfectly happy to accept that. Every family photo has at least one animal at the centre of it and if there was spare cash it was far more likely to go on a new horse rug, headcollar or dog bed than any human luxury.

This was my norm and I didn't realise until I went to school that not everyone lived like this. To be fair, literature played its part in asserting the dominance of animals – certainly the books I read, such as *The Jungle Book*, *Black Beauty*, *Doctor Dolittle* and *Winnie-the-Pooh* – but in real life, when I found that some people actually existed without regular animal contact, I was shocked.

I was so obsessed with animals that I went through a stage of thinking I *was* a dog. I often tell schoolchildren that dogs have got it sussed when it comes to the priorities in life: food, exercise, sleep and love. That's all that matters. I always think that if we humans stuck to these four essentials, we might have more chance of finding the key to eternal happiness. In the end, as dogs will tell you, everything else is gravy.

Me as a baby and Candy the Boxer.

Animals bring out the best of us as human beings and I think they define our humanity as well as our development as a civilised race. We have moved from being a type of animal to being dependent upon them, from using them as food or transport to treating them as part of our lives and our households. They reflect the best and the worst of us. If we are kind and consistent, patient and clear, they will respond by helping us as best they can. If we are cruel and impatient, they have every right to bite or kick us.

In the UK, around twelve million households have pets. That's millions of dogs, cats and rabbits as well as gerbils, hamsters and guinea pigs. Add in over 800,000 horses and ponies and you have a nation packed to the brim with animal-lovers.

Our relationship with animals as part of our lives and our households goes back much further than you might think and it seems to me that valuing the contribution of animals is a mark of an advanced society. Whether it's the ancient Mesopotamians, Egyptians, Native Americans, Greeks, Romans, Mayans or Incas, all the major civilisations of the world needed animals to progress and learnt to domesticate them. In China, the zodiac is made up of twelve animals and every year is guided by one of them. Your year of birth is linked to an animal, which is meant to exert influence over your character. I was born in 1971, the year of the pig. As such I am supposedly broad-minded, friendly, brave and kind, but apparently I can be a little lazy and not positive enough. From now on I will keep an eye out for any slack or negative behaviour and chastise myself accordingly!

In ancient times, many people worshipped animals as gods. In Ancient Egypt, the god Sobek, depicted either as a man with a crocodile head, or a crocodile, had to be kept sweet because of his associations with the River Nile, the lifeblood of the country. Most of the other Egyptian gods were a mixture of animal and human characteristics: Horus had the head of a hawk as god of the sky, Sekmet the goddess of war had the head of a lioness. Look up at the constellation at night and you will see the permanent reminder of our belief that animals occupy a higher standing – bear, scorpion, ram, eagle, crab, dog, swan and lion, all represented as shining stars above us.

The earliest myths, stories and poems celebrated animals and tried to explain where they came from or why they looked the way they did. One legend of ancient Rome describes baby bears being born as shapeless blobs, licked into the correct bear-like shape by their mothers, which is where we get the phrase 'lick into shape'.

The major religions teach respect for animals, and Buddhism places them on an equal footing with humans. It is highly symbolic that in Christianity, the son of God is born in a manger amid lowly farm animals and that a donkey carries him on his final journey. Christianity's patron saint of animals and ecology is Francis of Assisi (born 1181). Renowned for his love of animals, he was said to have preached to the birds and convinced a lion to stop attacking people and livestock. He looked out for any animal that was trapped or in need.

In this book, I wanted to find a way of celebrating all sorts of animals from giraffes and rhinos to cats and dogs, rats and chimpanzees to horses, sheep and pigs. It is a history of sorts, an examination of culture, art, sport, warfare and modern life. It is more than a collection of individual animals that have achieved amazing feats or added to our lives in some way – it is a tribute to all of them. A homage to their intelligence, loyalty, bravery, kindness and beauty.

There are all kinds of stories here: happy, sad, inspiring and funny. These animals are heroes to me – whether they've risked their lives to pull people out of the ruins of a bombed building, won a race against impossible odds, danced their way to an Olympic gold medal or been so incredibly badly behaved that they have managed to inspire something good (Stoffel, the honey badger, I'm thinking of you).

There are ship's cats and even a ship's pig. There is a Yorkshire Terrier found in the jungle who ended up saving a battalion. There is a hippo who walked the length of South Africa and a giraffe who walked across France. Christian the lion went from the Harrods pet department to running his own pride in Africa, while Wojtek the bear became a military hero in Poland. I have featured a talking seal, complete with a New England accent, life-saving canaries, empire-saving geese and dogs who can sniff out disease.

Some stories are personal, such as Thandi the rhino who I saw while filming on a game reserve in South Africa. I met the vet who saved her life after she had been mutilated by poachers who sawed off her horn. Having survived the attack, the operations and the skin grafts, Thandi has become the figurehead of rhino conservation across Africa.

There are a few equine heroes that marked my childhood – Aldaniti who won the Grand National in 1981, Milton who was my favourite show jumper and Charisma who I believe was the greatest eventer we've seen. From more recent times, Valegro makes the grade for his beauty and elegance in the dressage arena. All of them have widened the interest in equestrian sport with their brilliance and their personality. People just fell in love with what they saw.

The stories that most stand out for me are those of animal bravery. Take Roselle the guide dog, who guided her blind owner through the blazing destruction of the World Trade Center on 9/11. People were screaming, running and falling all around them, but thanks to her careful dedicated work her owner survived. Or the donkey Gallipoli Murphy who carried wounded soldiers from the battlefield; the pigeon Cher Ami who flew through heavy enemy fire to deliver a desperate message; and Sefton, the horse who survived the IRA bomb attack in Hyde Park and became a symbol of hope that, one day, the Troubles might be over.

Dogs, horses, pigeons, elephants, dolphins and many more have done remarkable things in the face of danger. Many of these animals have been injured or even died in the process. Some of their heroics are dependent upon natural instinct and a superior sense of smell, sight or hearing. Others are built on a quality that we value highly: loyalty. Some will argue they are victims rather than heroes, but I would rather appreciate and praise them, recognise and thank them for actions that may have been instinctive but which saved so many lives.

Many of these tales, even reading them back, leave me in tears because of the faithfulness on display, never wavering when we humans would have thought all hope was gone. Most dogs, for instance, are more loyal to their owner than their own partner, parents or children. They love completely and without compromise. The story of Hachiko waiting every day at 3pm at a station in Tokyo for his dead master who would never return had me in tears. For nine years, he continued to turn up on the dot of three. He became a symbol to the Japanese of eternal fidelity. Similarly moving is the story of Greyfriars Bobby, the Skye Terrier who, after his master's death, guarded the grave for fourteen years.

*

When Archie the Tibetan Terrier arrived in our house, I was convinced that his ancestral roots in guarding Buddhist temples would give him a strong sense of devotion and a certain zen. I hoped he would be eternally calm yet curious, have hidden depths and an ability to judge the character of a person just by smelling them. It turns out that a lump of rock had more zen than Archie, who would bite if his territory was invaded or if he was told to do something he didn't want to, who didn't much like bigger dogs, and who we never trusted with children. Archie was only interested in one thing: food. If bread was a forbidden substance, he would have been the best detective dog in the business. He was never perfect but we loved him dearly and he was a part of our family for over fifteen years, taking the best position on the sofa or the bed. He took us to places we would never otherwise have been, brought us friends for life and made us smile every day. Now he's gone, there is a gaping, dog-shaped hole in our lives and all we have are the photos (he was a very photogenic dog) and the memories.

There are dogs who can predict the onset of an epileptic fit or detect cancer, find missing pets or sniff out explosives. Archie could smell a dropped crisp at fifty feet. I think he would have found his way to the front door if left some distance away but not if someone else offered him a decent meal on the journey, and he certainly wouldn't have been able to do what Bobbie the Wonder Dog did when he travelled 2,500 miles over six months to find his way home.

I have learnt so much through writing this book about the physiology of different species of animals and I hope you will enjoy finding out about their brain power or the fact that dogs have 300 million olfactory receptors compared to our six million, which would explain why they can smell so much more than we can.

There are some things that science doesn't have an answer for – such as how homing pigeons and migrating birds can find their way home or why starlings fly in murmurations of thousands – but that's when I'm happy to be lost in wonder at the sheer brilliance of the animal kingdom.

As Archie made us even more aware, there is a deal we make when we allow ourselves to love an animal – there is a high probability that he or she will die before us and we will experience huge pain. Alice and I were in bits when Archie left us, and we cried for days. I still find it difficult, especially when it's bedtime, but I would rather suffer the grief of a broken heart than never have known the joy of having Archie in our lives.

Some animals not only bring joy but give the lives of those around them a whole new meaning and purpose. Bob the cat changed the life of James Bowen, offering him comfort as well as the inspiration to write a best-selling book. Wheely Willy gave hope to children with his joyful attitude to life, despite being paralysed by the cruelty of his original owner, while Magic the miniature pony has provided solace to children and adults recovering from trauma.

If I asked my father to name a heroic animal, he would not hesitate. The name Mill Reef would be instantly on his lips because the horse he trained to win the Derby in 1971 changed his life. Mill Reef was an outstanding racehorse – a champion at two years old and the winner, at three years old, of the major Group 1 races in Europe – the Derby, the Eclipse, the King George and the Prix de l'Arc de Triomphe. He was brave and talented, seeming to float over the ground like a ballet dancer. He was duly celebrated as a superstar, but the following year on the gallops at home, he broke his leg.

Dad was distraught. Operations to mend a broken leg for a horse are fraught with danger. The anaesthetic process can be fatal if the dosage is not spot on and the horse can do itself more damage when it comes round by thrashing out in a panic. There is also the difficulty of keeping the weight off a broken leg without the horse developing other issues because of lack of movement. Luckily, Mill Reef had the temperament and the constitution to withstand the veterinary procedure and the recovery. That, for my Dad, is why he is a hero. He never raced again but he went on to have a very successful career as a stallion and there are still horses descended from Mill Reef winning races today.

My Mum would probably pick Candy, her first Boxer, who was so attached to myself and my brother that she threw herself out of a top-floor window when she thought we were being kidnapped. We weren't – Mum was just wearing a new coat and Candy didn't recognise her from the back – but my word, it was an impressive act of protection.

I mentioned earlier the ponies who taught me how to behave. Valkyrie was a funny little ball of fur who had seen the upper end of life as a Shetland pony. She had belonged to Her Majesty the Queen and had been given to my parents when I was born. Valkyrie expected certain standards. She did not take kindly to a toddler throwing a tantrum or slamming down a dandy brush in anger. She had a way of looking at me with disapproval if I behaved badly and, on occasion, would gently but persistently back me into the corner of the stable, refusing to let me out until I got my little temper tantrum under control.

Me in rock-star mode on Valkyrie the Shetland, my brother Andrew sitting on the grass.

It was from the back of Valkyrie that I started to fall off and actually to enjoy the sensation of unexpected departure. I learnt to fall and roll, and to let myself relax in the moment of falling rather than putting out a stiff arm, which could then get broken. Learning how to fall off before you learn how to stay put may not be the way every child learns to ride but it worked for me.

As I got older and taller, I moved on to bigger ponies. Volcano was a very pretty Welsh Mountain who had a naughty streak. I think he used to stop or run out at jumps on purpose to make sure I was concentrating. Every time I got a bit overconfident or my concentration lapsed, he would jam on the brakes. It was infuriating but educational. Looking back, I think he taught me to keep my focus and to be patient, to keep going even when things are looking bad and, most of all, not to get too full of myself.

All the while, I was trying to find a way to communicate with animals. I really wanted to be Dr Dolittle. I wanted to understand everything they were saying and thinking. Maybe I was over-thinking the relationship but I believe my favourite pony, Frank, understood me.

In my eyes, Frank was a beauty because he was so original. His coat was mucky grey with brown and black splodges in random places, his ears were brown, the pink skin around his eyes and nose was liable to burn in the sun, and his mane was constantly rubbed into a spiky mess. He had a tendency to charge out of the stable and once broke my foot by treading on me as he made his escape, and he was not an easy ride. He had a mouth about as sensitive as a block of wood so steering was an issue, but I loved him.

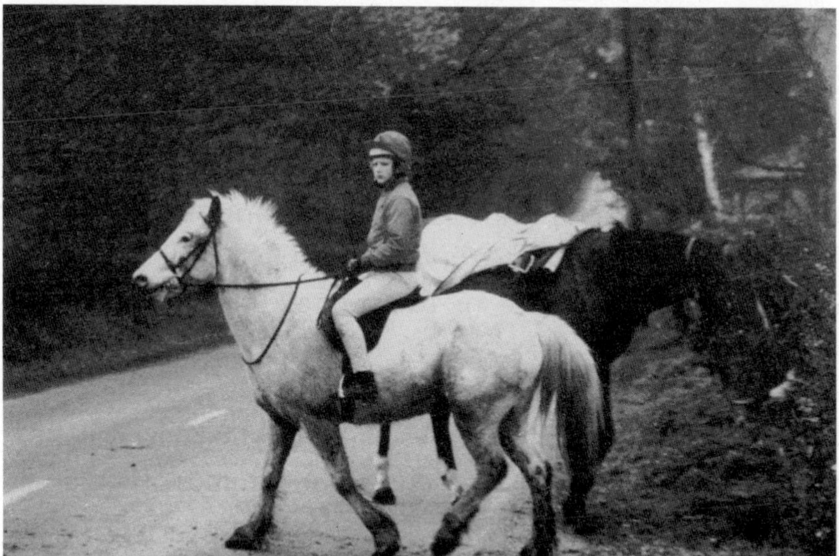

Me on Frank, my first love.

He was my Frank and I would ride for hours telling him all about my problems, worrying out loud about how I didn't fit in and working through issues that bothered me. Frank listened. He really did. His brown ears flickered back and forth and I felt better talking about things out loud. I looked at him and admired his confidence, his refusal to be defined by conventional descriptions of what a pony should look like, his enjoyment of life, and I decided that Frank knew that it was OK to be different. In fact, he embraced his difference and celebrated the fact that he didn't look or behave like other ponies. He would always stand out and I loved him even more for that.

Henry was a different matter. He was very handsome and very, very fast. Henry didn't do anything at a sedate pace. He set off at a hundred miles an hour and didn't slow down for anything. I could just about keep him in a collected canter in a show-jumping arena but as soon as he was in an open space, he was off. Riding him cross-country was one of the most terrifying and exhilarating things I've ever done. I had to walk the course really thoroughly because if I didn't know where I was going next, we'd be past the point before I had worked it out. I have never jumped fences at that sort of speed, not even on a racehorse. I had to learn to feel the fear and do it anyway.

That may have helped when it came to riding in races. For the short time that I was an amateur jockey I realised that time could actually slow down. I would feel nervous and tense before the race and then, as soon as I got the leg up and was in the saddle, it was as if I had entered a trance. I wasn't nervous any more and, during the race, I saw things about to happen and managed to respond before they'd even taken place. I guess it's what they call being 'in the zone' and it's a blissful state. It showed me how to be fully present, actively alert and yet completely calm.

I always thought it was difficult for a jockey to have a real connection with a racehorse. They're perched up their necks with little contact through the leg and they're either pulling hard against them to slow them down or pushing with all their might to make their horse go faster. There isn't much in between. Unlike with eventing, show jumping or dressage, a jockey doesn't spend years getting to know a racehorse. Sometimes, they may hop on them for the very first time at the races. It seemed to me a bit like driving a car – the makes might be slightly different but the controls are always in the same place.

It was only when I started to ride a horse called Knock Knock that I realised racehorses are not all the same. He had talent, there was no doubt of that. At home on the gallops he could outrun any other horse but on the racecourse, he didn't want to know. He had been unplaced so often that his form read like a string of duck eggs: 0000. I started to ride him on the gallops at home and noticed that he really hated other horses. He would put his ears back and try to bite them if they came too close so I kept him away and gave him his own space.

He was very responsive to affection and loved to be patted. As we walked along, I kept patting his neck and he would prick his ears, while in the stable I gave him lots of cuddles and polo mints. I really enjoyed riding him and when we were allowed to do a bit of work with a horse who had won his last race, Knock Knock cruised past him as if he were standing still.

My father decided to enter Knock Knock in an amateur race, just to do something different. He asked me if I wanted to ride him and I couldn't say yes fast enough. It was a conditions race in which the conditions were not at all in Knock Knock's favour. He was carrying more weight in relation to the other runners than he would in a handicap and, not surprisingly, he was an outsider in the betting at 25/1. We cantered down to the start with his head tucked in beautifully like a dressage horse. I thought that whatever happened, at least we looked good.

When the starting stalls opened, the other runners set off at a furious pace. Knock Knock and I took our time and I let him find his stride, keeping away from the pack and letting him have a bit of space. As we turned into the straight, the front runners were coming back to us and Knock Knock was cruising. There was no pressure on us and I just wanted him to enjoy himself so I gave him a pat on the neck and said, 'Good boy'. He suddenly took off with a burst of acceleration that took us sweeping round the whole field. We passed the second favourite, the favourite and suddenly, with just under a furlong to run, we were in front. I kept saying 'good boy, good boy' and laughing at the ridiculousness of it all. I didn't move in the saddle and I certainly didn't pick up my stick. I was grinning all the way to the winning post as we came home in front.

I rode Knock Knock a further twelve times, by far my most consistent partnership with a racehorse, and won another three races on him as well as finishing placed, eight times. It was so much fun because it was as if I'd found a way to solve a riddle and really had developed a relationship with him that meant he wanted to win for me.

There is a nobility about horses that has inspired artists over the centuries to try to capture their dignity as well as their beauty. Look at the Munnings portrait of Warrior, the cavalry horse who survived the horrors of the First World War, and you see the way his head is turned away from the artist, looking out for danger rather than wanting attention. He is proud yet humble, alert yet steady. Sir Alfred Munnings was one of the greatest equestrian artists of all time and his skill is in portraying the character qualities as well as the physical presence of horses.

Across the world, horses have always been highly prized for their strength, speed and agility. A means of transport, a partner in hunting or a comrade in war – as such they were a game-changer for our ancestors. There are hundreds of statues of horses around the world, showing their status and the significance we bestow upon them. In early graves, the powerful have been found buried with their most precious possessions including their horses.

In Charlie Mackesy's brilliant book *The Boy, The Mole, The Fox and The Horse*, it is the horse who offers wisdom and comfort. In every situation he knows what to say without seeming didactic or trite. The boy asks him to define courage and the horse says, 'To tell the truth of who you are with your whole heart'. The horse teaches the boy about the power of love, about not looking too far ahead, about asking for help and not being swayed by other people's behaviour. The horse is direct, honest and unfailingly kind. He represents everything we wish to be and yet he doesn't have an ounce of arrogance. 'The truth is, everyone is winging it', he says in a powerful message to all of us with imposter syndrome.

It is no surprise that horses are proving so valuable in a range of therapy treatments. Whether it is working with those who have gone through unspeakable trauma, those fighting alcohol and drug addiction, those with physical challenges or neural diversity, those who struggle at school or who are trying to adapt to a new life, equine therapy has been a successful and highly valued treatment. There is something about horses that speaks to our souls, and they see in us something that often we cannot recognise in ourselves.

Dogs have also proved a big success in the therapy sector and even the odd cat, as Thula has proven, will reach the parts that fellow humans cannot access.

We all have our own animal champions and I suspect that in times of uncertainty, such as the 2020 months of lockdown, it is the everyday heroism of our pets that keeps many of us going. They look after us as much as we look after them. For those living alone they provide company and comfort, a reason to get out and walk every day. For those living with others they can be the family focus, the magnet to ensure you all do things together. For those living in dangerous circumstances, they bring relief and hope. As Herman Melville wrote, 'No philosophers so thoroughly comprehend us as dogs and horses.'

It is telling stories about our beloved pets that keeps them alive for ever and that is what I would like this book to do for Archie and these 100 exceptional, extraordinary, remarkable animals. I hope you enjoy reading it as much as I have enjoyed writing about them.

ALDANITI

The Grand National is the most famous steeplechase in the world. Grand because of the size of the fences, national because of its grip on the nation, it commands audiences of over 500 million worldwide. It takes a horse of stamina, courage and excellent jumping ability to win. First run in 1839, when it was won by the aptly named Lottery, the race has produced some of the greatest sporting stories in history, but I believe none is more inspirational than that of Aldaniti and his jockey, Bob Champion.

I was ten years old in 1981 and I remember the build-up to the race vividly. Champion had just come through surgery and six months of treatment for testicular cancer, and many people thought he wasn't yet fit enough to ride in what is the longest race of the season. Aldaniti had his own fitness problems. His legs had always been fragile, so much so that one vet said he would never be able to tolerate a racing career.

While still undergoing his chemotherapy, Champion went to watch Aldaniti run at Sandown in November 1979, with the notion that the two of them might be able to team up in the 1980 Grand National. His hopes were dashed when the horse finished lame. Aldaniti went to his owners, Nick and Valda Embiricos, to rest and recuperate. Bob went back to hospital. Both of them were at their lowest points.

After he'd had a year off, the horse's trainer Josh Gifford

brought Aldaniti back into training, with running in the Grand National as the goal. Both the Gifford and the Embiricos families were conscious that Champion was looking forward to riding the horse again and they didn't want to let him down. No one knew whether that would be possible but at least in Aldaniti they could keep that dream alive. 'They gave me six to eight months', says Champion,

> or a 35–40 per cent chance of living, with the treatment. Josh always said my job was there. I know he never really thought I'd live, but he kept giving me the confidence, which I needed.

The pair were reunited for a handicap chase at Ascot in February 1981. It was the equivalent of a reconnaissance mission for both of them. Not much was expected but Aldaniti had other ideas. His legs may have been weak but his heart was where his real strength lay. He won by four lengths. Champion says of the chestnut gelding, 'He absolutely bolted up. He never came off the bridle. I thought – if he's as good as that on the day, that's good enough for the National in my book.'

And so it was that Champion and Aldaniti made it to Aintree on 4 April, starting at 10/1, second favourite behind Spartan Missile.

The race didn't begin well, Aldaniti almost coming to grief at the first fence. He stood off too far on take-off, landed too steeply and his nose grazed the turf. He wasn't much better over the second but Champion sat tight and, by the third fence, they found their rhythm. After that, he jumped superbly and came to the last of the thirty fences full of running, with the favourite Spartan Missile also going well.

Aldaniti and Bob Champion winning the Grand National at Aintree in 1981.

No one watching on television will ever forget Sir Peter O'Sullevan's commentary, his gravelly tones rising to a crescendo as the two horses entered the final furlong:

> It's Aldaniti in the lead but being pressed now by Spartan Missile. It's Aldaniti from Spartan Missile and here comes John Thorne, fifty-four-year-old John Thorne is putting in a storming finish. It's Aldaniti from Spartan Missile. Aldaniti is gonna win it, at the line, Aldaniti wins the National!

John Thorne, a grandfather who had also owned and trained his horse, would have been a decent enough story in its own right but Champion and Aldaniti had both come back from near-death experiences, both had been written off, both had triumphed against the odds, both were inspirations to millions. It was a dream outcome. Watching at home with my brother and parents, I cried. Mum knew Bob well, Dad was a friend of Josh Gifford. We were all deliriously happy for them. Far outside racing circles, people were touched by the story and inspired by the outcome. 'When I went by the winning post in front, I did think at least it will give a lot of people in hospital hope', Champion reflected.

Aldaniti, named for his breeder Tommy Barron's four grandchildren (Alastair, David, Nicola and Timothy), returned to his Sussex stable a hero, with more than 3,000 people turning up to welcome him home.

Champion and Aldaniti returned to the Grand National in 1982 but fell at the first fence. That same year, the horse was retired from racing. The pair were immortalised in a book, *Champion's Story*, which was later made into a film called *Champions*, starring John Hurt as the jockey and Aldaniti as himself. There aren't many films about racing that translate to a wider audience, but this story had captured the nation's heart.

Champion set up the Bob Champion Cancer Trust in 1983, and Aldaniti spent much of his retirement helping to raise funds for research into the disease that almost killed his jockey. They raised millions and whenever I saw them together, the bond was obvious. Aldaniti was a natural showman and he loved a big occasion. In 1987, he walked from Buckingham Palace to Aintree ridden by 250 different jockeys, including HRH The Princess Royal. He received a hero's welcome as Bob rode the last mile to the racecourse on Grand National Day. That walk alone raised £820,000, and during his life Aldaniti helped raise over £6 million. 'I used to see him regularly. He did so much for the cancer trust', Bob said. 'He used to go up in the lift at big stores. Terrific temperament. He was wonderful with the public – a nice, kind horse who loved the attention.'

Aldaniti's active and happy retirement ended in March 1987, when he died of a heart attack at the ripe old age of twenty-seven. The nation mourned. A British Rail Class 86 electric locomotive that could reach speeds of up to 100 mph was later named in Aldaniti's honour, and Champion found his own way to remember the horse who had helped him fight his way back to life:

> The most expensive thing I have ever bought was a bronze statue of Aldaniti. It cost £3,000. I just love it. It reminds me of that day. I loved Aldaniti. In my opinion, winning a race is 98 per cent down to the horse. I always thought he would win a National.

ALEX

Parrots are the most intelligent animals on earth. They have managed to master language, can count, have displayed logic, have an excellent ability to mimic accents and can even remember a stand-up comedy routine. Put them in charge of a global pandemic and they might have found a way to solve the crisis. History is full of stories of talking parrots and their close ties with humans. They have been kept as pets in Europe since 327 BC, when Alexander the Great's forces conquered India and took them back home to Greece.

Talking parrots were so prized among the Roman upper classes that slaves were employed to teach the birds Latin. The *Kama Sutra* (written over a thousand years ago) stated that one of the sixty-three requirements for a man was to be able to teach a parrot to talk.

Many of the cleverest and best-known parrots have been African Greys. Not surprisingly, Polynesia, the parrot in the *Dr Dolittle* books who teaches Dolittle the language of the animals, is an African Grey. King Henry VIII famously owned an African Grey which, legend has it, amused itself by calling out to local boatmen to row across the Thames to Hampton Court Palace. The boatmen who rushed to obey the royal summons were not happy to discover their services were not in fact required. Hard to please, Queen Victoria was *finally* amused when her African Grey, Coco, was taught to sing 'God Save the Queen'.

America's very popular seventh president, Andrew Jackson, bought Poll, an African Grey, for his wife Rachel, but ended up taking care of the bird himself. Famously, Poll had to be removed from his owner's funeral. According to volume three of Samuel G. Heiskell's *Andrew Jackson and Early Tennessee History*:

> Before the sermon, and while the crowd was gathering, a wicked parrot got excited and commenced swearing so loud and long as to disturb the people and had to be carried from the house . . . where[upon Poll] let loose perfect gusts of cuss words.

I suppose we all deal with grief in different ways.

It is possible that it was Jackson, a former soldier, who taught the bird to swear like a trooper. They were certainly close enough; and Poll was Jackson's chosen portrait companion.

Scientists have long been fascinated by the parrot's ability to speak. Aristotle described it as 'the bird that is said to be human-tongued (that becomes more outrageous after drinking wine)'; whereas Pliny the Elder, the Roman scientist, suggested that the best way to teach parrots to talk was by hitting them on the head with an iron bar because they wouldn't feel it. (I would suggest that *he* had drunk too much wine when he said this.)

Scientific research shows that the way a parrot's brain is wired makes it possible for the birds to repeat human words, but it was assumed that their linguistic skills were just those of imitation. In English, we use the verb 'to parrot' to refer to someone who copies another's words and we use 'parrot fashion' as a way of describing the ability to remember something so well that we can repeat it verbatim. It was believed that you could teach a parrot to say 'hello', sing a song, or swear but that was about as far as it went.

It took a very special bird to prove that assumption wrong. In 1977, psychologist Irene Pepperberg bought a one-year-old African Grey from a pet shop. She wanted to investigate if a parrot could learn language.

Other research projects at the time were using animals with large brains, such as primates or dolphins. Pepperberg describes the scientific community's response to her initial grant proposal as

> underwhelming . . . They were basically asking me what I was smoking. They were horrified I was going to do this with a creature who had a brain the size of a shelled

> walnut [the human brain by comparison is the size of a small melon and weighs three pounds], a creature that was a pet. How was I going to maintain my scientific objectivity?

Eventually she was awarded a grant and was able to begin work in earnest. She called the parrot Alex (short for Avian Learning Experiment rather than plain old Alexander) and taught him words by letting him decide on his own reward, always connecting specific words to the same rewards. So by learning words, Alex gained more control over his surroundings. He could indicate which snacks he wanted as a reward and when he wanted to take a break or go outside.

In this way Alex developed a vocabulary of around 150 words and was able to recognise fifty objects. He could understand and answer questions about the objects. He learnt to recognise colours, shapes, materials and functions. For example, he knew what a key was for. He recognised new keys as keys, even if they were a different shape. He understood concepts such as 'same', 'different', 'bigger', 'smaller', 'yes', 'no' and 'absence'. He also registered numbers and could count up to eight. He knew how sentence structure worked and could combine words. When Pepperberg and her assistant made mistakes, Alex corrected them. He sometimes practised words when he was alone. Alex once asked Pepperberg what colour he was himself, a pretty existential question for a parrot.

Pepperberg and Alex worked closely together for thirty years. When she got divorced, her husband got custody of their dog and she kept Alex, describing him as 'my scientific colleague not my pet'. Pepperberg claimed that Alex was emotionally on the same level as a two-year-old human.

When he was bored, he sometimes gave the wrong answers on purpose or answered the questions incorrectly, despite knowing the correct answers.

The pair appeared on a number of documentaries together, and on one occasion were asked to record a show for BBC radio. Pepperberg was bemused – how was she going to show that whatever the parrot identified was correct if no one could see them? But she decided they'd give it a go all the same.

> Alex knew what buttons to press. We started with an orange toy. I took it in to Alex and asked, 'What colour?' He replied, 'No, what shape?' I said, 'OK, Alex, it's four-cornered but can you tell me what colour?' He answered, 'No, what matter?' I said, 'It's wood, Alex, can you tell me what colour?' He responded, 'How many?' I gave up and left the room, then heard a little birdy voice saying, 'I'm sorry. Come here. Orange.'

Alex's extraordinary contribution to the field of human–animal communication turned previous research on its head. It fundamentally changed the perception that parrots act only out of instinct. Alex's vocabulary remained limited, but his use of words and concepts showed real understanding and intelligence.

On 6 September 2007 Alex died suddenly and unexpectedly. Pepperberg was devastated: 'I realised I had lost the most important being in my life for the last thirty years.' His last words had been the same words he spoke to her every night before she left the lab: 'You be good. I love you. You'll be in tomorrow.'

ANTIS

This is the tale of how a puppy abandoned in no-man's-land grew wings and flew into aviation history.

On a bitterly cold morning in January 1940, an Allied plane on a reconnaissance mission over enemy lines was shot down by the Germans. Czech airman Robert Bozdech managed to scramble free, helping injured French pilot Pierre Duval out of his harness and dragging him clear of the wreckage. Resting in a snow drift, Bozdech looked around for somewhere safer to take cover.

In the distance, he could see a farmhouse. He decided to make a recce, but when he reached the building, he found that the place had been ransacked. Apart from a dusty table, some logs and a frying pan on the stove, there was little left. But then came a faint scratching and a muted whimper. Pistol drawn, he advanced cautiously towards the source of the sound. 'Show yourself', he commanded. He could hear breathing, a slight snuffle, but no one came forth. He tried again. Still nothing. Moving forward, ready to shoot, he suddenly spotted his 'enemy' – a small German Shepherd puppy. He picked him up and tucked him into his jacket to keep him warm.

That evening, Bozdech and Duval talked through what they needed to do to get back behind Allied lines before German forces resumed their search in the morning light. But making a break for it was very risky, even without a dog in tow. They had no choice but to leave the pup with food and water, and head out to take their chances without him.

Almost as soon as they left the house, the sky grew bright from flares lit by the Germans, who were still searching for the downed enemy crew. As Bozdech dragged Duval through the snow to the safety of some nearby trees, he heard the dog's pitiful howling, threatening to blow their cover. There seemed to be little choice but to go back and kill him.

But one look at that pleading face, those soul-searching eyes and he just couldn't do it. Instead, he tucked the puppy into his flying jacket and went back to Duval. Somehow all three made it into the woods where they were rescued. Duval was taken to hospital, while Bozdech was flown back to base at Saint-Dizier, still clutching the puppy in his arms.

The new arrival was a hit with his fellow soldiers, who lavished him with affection and food. But he needed a name. Bozdech and the other exiled airmen decided to christen him in honour of their favourite aircraft, a Russian Pe-2 dive-bomber known by the Czech air forces as the ANT.

Bozdech and Ant, later renamed Antis, became inseparable, the dog barely leaving his new master's side and obeying his every command until one day he refused to come when called, instead standing rigidly in the middle of the base with his eyes fixed on the horizon. Reliable as any radar, he had anticipated the incoming German planes that unleashed a two-hour bombardment on the base. He later alerted them to another aerial attack which wreaked havoc. The destruction was so bad that Bozdech and Antis got separated in the chaos. When the dog was nowhere to be found, Bozdech grew ever more frantic, imagining the worst. Three days later, Antis reappeared, badly injured. He had been blown up and covered in debris but had somehow scratched his way out and survived. When he went on to make a full recovery, he'd gained a reputation as the 'radar dog'.

As the situation in France deteriorated the Czech airmen made their way south, heading towards Spain and then Gibraltar so that they could travel to the UK and continue their fight from there. The journey was difficult and dangerous, with the men taking turns to carry the dog on their shoulders. They took an overcrowded train to Marseille which made slow progress, moving only sixty miles in three days. At one stop, the soldiers jumped off and attempted to milk a cow in a nearby pasture and fill a baby's bottle for the dog. The locals assumed they were after milk for a baby so gave them precious supplies.

They finally reached Gibraltar, intending to travel on the MV *Northmoor*, part of a convoy heading back to Britain, but there was a problem. Guards refused to allow animals onto the ferry for the cargo ship. Bozdech could not abandon Antis now. Taking a chance that the dog would do anything to stay with him, he left Antis sitting obediently on the shore, and then, when he was safely on board the *Northmoor*, climbed down to the swimming platform and called for him. Antis swam out 100 yards to be scooped out of the sea and smuggled into the hold in Bozdech's greatcoat.

It was an eventful journey, a submarine attack followed by an aerial one with the resultant damage meaning a transfer to a new vessel, the *Neuralia*. The Czechs looked to smuggle Antis between the two ships by hiding him in a kitbag. This failed when the dog pushed his head out of the bag just as they were boarding – fortunately the new crew seemed happy to have him on board. But with no money to pay United Kingdom quarantine fees, Antis found himself being illegally smuggled once again, this time in a bag of cargo.

Bozdech joined the RAF No. 311 (Czechoslovak) Squadron based in Liverpool, where he had a frustrating desk job. Still, Antis was a hero. He warned Bozdech of an incoming air attack and then helped to recover six surviving victims of an air raid, sniffing them out and clawing through rubble to help free them. His own paws were injured in the process but Antis wouldn't give up.

Bozdech was eventually given a flying job again at RAF East Wretham near Thetford in Norfolk. Regulations prohibited Antis taking to the air with his owner, but the dog had other ideas. As the airmen were preparing for an aerial assault on northern Germany, he vanished. Bozdech was beside himself but had no option but to carry on with his mission. At 16,000 feet, when they were about to begin bombing, he discovered Antis hiding in the belly of the plane, gasping for air. The two took it in turns with the oxygen mask. Bozdech's pilot thought him mad but it was only the beginning of Antis's extraordinary flying career.

Antis with Robert Bozdech.

The dog became the squadron's mascot, wearing a specially crafted oxygen mask on every sortie. He sustained injuries twice from gunfire to the plane but both times Bozdech didn't even realise until they landed because Antis never made a fuss. 'He didn't whine, he didn't panic. He showed courage that perhaps a human being couldn't show', Bozdech said.

Together, they completed around thirty missions.

After the war, Bozdech returned to his native Czechoslovakia, taking his faithful friend with him. It was not a happy homecoming because of the newly powerful Communists, who were determined to persecute anyone who had served on the side of the Western Allies. Bozdech was once more forced to flee and Antis helped him escape, avoiding gunfire and searchlights to cross safely into West Germany and then to the UK. There, Bozdech became a British citizen. Antis remained with him and, in 1949, was awarded the Dickin Medal, the highest British honour accorded to animals, in recognition of his heroic war work.

After the pair had spent thirteen happy years together, the faithful German Shepherd died peacefully one night. Although he loved all dogs, Bozdech could never have another. Antis was irreplaceable, his one and only.

The PDSA Dickin Medal for gallantry is recognised worldwide as the animal equivalent of the Victoria Cross and is the highest award any animal can receive while serving in military conflict. It was instituted in 1943 by Maria Dickin, founder of the PDSA (the People's Dispensary for Sick Animals), inspired by the 'devotion to man and duty' shown by animals on active service, and acknowledges devotion to duty or outstanding acts of bravery shown by any animal serving with either the armed forces or civil defence units in wars across the world.

To date the medal has been awarded seventy-one times (plus an honorary award in 2014). Thirty-four dogs have received the honour, along with thirty-two pigeons, four horses and a cat.

ARAMIS AND BROTHERS

Drones are one of the great and growing threats to modern security. While they undoubtedly have many legitimate uses – including surveillance and research – these flying robots, now easily bought online or from toyshops, can be modified by terrorists for nefarious means.

So, how do you take down a flying spy or a tiny, remote-controlled airborne weapon? You could shoot it but that has the potential to cause chaos, especially if the drone, with rotors still turning, were to fall into a crowded area. In the war against unauthorised flying objects, it makes sense to use a superior flying object with a brain and a sense of direction. Enter the king of all birds of prey: the eagle.

In 2015, a wild wedge-tailed eagle was caught on camera attacking and disabling a drone in Australia. The previous year researchers from the Royal Melbourne Institute of Technology had launched a glider via autopilot to mimic the slope and soaring of birds. This too was taken down by an eagle. The instinct to attack and destroy a drone seemed inbuilt in this superhero of a bird.

As part of an experimental programme run by the French military, four precious Golden Eagle eggs were hatched on top of drones, and kept there as they learnt to feed. With a wingspan of six to seven and a half feet, a weight of up to fifteen pounds, vice-like talons and the ability to dive at a speed of 150 to 200 miles per hour, the Golden Eagle is a naturally fearsome warrior. All that was required was to direct that aggression towards the correct target on instruction.

The birds were named after the musketeers of Alexandre Dumas fame, Aramis, Athos, Porthos and D'Artagnan, and were taught to associate UAVs – unmanned aerial vehicles – as prey and take them down on sight. For this very twenty-first-century skill they were rewarded with meat, as falcons would have been in medieval times.

But could these golden avian musketeers really be used to destroy drones invading their airspace? At the French air-force base in southwestern France, a trial flight showed D'Artagnan attacking a drone and, after covering 650 feet in just twenty seconds, bringing it down to the ground with force. As the air-force general explained to Reuters news reporters, 'These eagles can spot the drones several thousand metres away and neutralise them.'

As we know, personal protective equipment is essential, even for superheroes, and with that in mind, these soaring, powerful creatures now have specially designed mittens of leather mixed with an anti-blast material to protect their talons.

ARKLE

'Arkle, Arkle, Arkle, Arkle' was what everyone in Ireland was talking about. 'Arkle for President', read grafitti on a Dublin wall. In a nation where horses are idolised but most of the good ones sold to the UK, the racehorse Arkle became a national treasure, a symbol of Irish pride, and the embodiment of courage and sporting excellence.

Along with George Best, Rod Laver, Pelé and Muhammad Ali, Arkle was one of *the* sporting icons of the 1960s. He was so good, they had to rewrite the rules of racing to give other horses a chance.

But what was it about this good-natured bay gelding that made him a living legend and saw stamps, tea towels and all manner of other memorabilia emblazoned with his image? What was it about a horse who cost them money that prompted bookies to doff their hats to him? Letters addressed simply 'Arkle, Ireland' found their way to his stable. With an almost religious reverence, the horse became known as 'Himself'. Admiringly called a 'freak of nature' by Sir Peter O'Sullevan, Arkle's fame stretched far beyond the world of racing.

Foaled in 1957 in Co. Dublin, he was bought for 1,150

guineas as an untried three-year-old by the Duchess of Westminster. A keen horsewoman originally from Ireland, she was a sporting and enthusiastic owner of steeplechasers. She named her new purchase after a mountain on her Scottish estate in Sutherland.

The Duchess kept Arkle in Ireland and sent him to Co. Meath to a trainer with whom she had enjoyed great success – Tom Dreaper, who had also trained Arkle's dam. Both Anne, Duchess of Westminster and Dreaper shared a quality that worked in their horses' favour: patience. Arkle ran in a couple of bumpers (flat races for horses bred for jumping) without showing much flair. He clearly couldn't see the point in racing if there was nothing to jump.

As a five-year-old he made his debut over hurdles. He was sent off at 20/1 by the bookies, disregarded and unfancied. He won easily and suddenly Dreaper knew he had a good horse on his hands. When he started jumping full-size fences, he was even more impressive. By the time Arkle had won the Gold Cup at Cheltenham three years running, two Hennessy Gold Cups, the King George VI Chase, the Whitbread and the Irish Grand National, carrying two stone more than his rivals, it was clear Dreaper had trained a great horse. In fact, many still argue that Arkle is *the* greatest jumper ever seen.

The Duchess of Westminster argued that it was his personality as well as his winning record that won him a legion of fans. With his head held high, ears pricked, he seemed aware of and interested in everything around him and enjoyed the attention. 'He always looked so terribly proud and rather showing off, almost playing to his public', she said. 'It's always been the same and I think he loves people looking at him.'

Arkle wins the King George VI steeplechase at Kempton Park in 1965.

He was docile in the yard and children could pet him. He was kind as well as intelligent. There were stories aplenty about Arkle. A photo was printed of him drinking stout and so Guinness jumped on the PR bandwagon and sent him a lifetime's supply. People joked that he got his strength from a daily diet of oats, raw eggs and Guinness.

His Timeform rating (the racing equivalent of school grades or Michelin stars), at 212, was the highest ever awarded to a steeplechaser. He galloped with ease and snapped his knees up high when he jumped a fence, crossing his legs slightly in front before landing. He was strong without being overbig, accurate in his jumping without being over extravagant and never fell once. He won twenty-seven of his thirty-four races, including four of his five meetings with the giant Mill House, regarded as the best horse trained in Britain at the time.

Together with jockey Pat Taaffe, Arkle dominated National Hunt racing in the mid-sixties, largely thanks to his exceptional powers of acceleration that led one leading jockey to remark, 'The b------ went past me as if I were a double-decker bus. He'd have won with the whole Taaffe family on his back', while another added, 'There wasn't anything I could do. I'll swear he was laughing as he passed.'

The handicapper for the Irish Grand National had to devise two weighting systems, one to be used if Arkle was running and one if he wasn't – an unprecedented move which meant that even though he won the 1964 race by just one length, he did so carrying two stone more than any of his rivals.

After victories in the 1964 and 1965 Gold Cups at Cheltenham, he was the shortest-priced favourite in history to win the 1966 prize. Despite a mistake early on, where he ploughed through a fence because he was looking at the crowd, he won the race by thirty lengths. Arguably his best career performance was in the Gallagher Gold Cup at Sandown in 1965 when he beat Mill House by thirty-two lengths, giving him sixteen pounds in weight *and* breaking the track record by seventeen seconds. That record still stands.

Arkle's last race was in December 1966, in the King George VI Chase at Kempton Park, where he struck the guardrail with a hoof while jumping the open ditch and fractured his pedal bone. Despite the injury, he still managed to finish second, beaten barely by a length. He was only nine years old.

He was operated on and the foot put in plaster for six weeks. When he returned to Ireland, he was rested and hopes were high that he might make a comeback. Get Well Soon cards, presents and good luck mementoes arrived for him. A full-time secretary was required to answer all the post.

Sadly, he never fully recovered. He was officially retired in 1968 and went to live on the Duchess of Westminster's farm in Co. Kildare. In 1969 he came to the Horse of the Year Show in England to take part in a Parade of Personalities. He loved being the centre of attention once again, but by the following year arthritis had set in and he was struggling to walk. The Duchess had to make the decision to have him put down. She wept as he went to sleep for ever and the grief was felt all over Ireland as their great hero breathed his last.

Five years after he had been buried, the Duchess had his body exhumed and his skeleton is now on display at the Irish National Stud museum in Co. Kildare. There is a statue of him at Cheltenham Racecourse and in April 2014 a life-size bronze statue was unveiled in Ashbourne, Co. Meath. It was deliberately set in a public space rather than a racecourse so that everyone could see him and pay tribute to his achievements, his status and his impact on the cultural and sporting history of Ireland.

BARRIE

This is the story of how a tiny puppy, rescued from the rubble of war-torn Syria, saved the life of the man who found her.

From his first day as a soldier, Sean Laidlaw found himself

in the thick of the action. The work was dangerous and gruelling, 'It started to sink in that I could die at any moment.' Over the next ten years with the Royal Engineers, he witnessed many atrocities. He dealt with this in what he thought of as 'the British way': he kept a stiff upper lip and cracked a few jokes. That changed one day in Afghanistan, when he came across the body of a British soldier who had been brutally tortured by the Taliban. 'I blacked that out for a long time', he explains. 'It went into a box in my head . . . until that box ended up opening for ever.'

He found it really difficult when people back home, with gruesome curiosity, asked whether he had killed anyone. When his partner suffered a miscarriage he found he couldn't cope at all: 'I was angry at the world, at everyone for any reason whatsoever.'

His relationship sadly disintegrated, and with it went his home. Laidlaw turned to exercise, in theory to give his days structure but in reality to punish himself. He pushed his body to the limit three times a day and started to pump steroids to bulk up his muscles. His life was spiralling out of control and Post Traumatic Stress Disorder had taken hold.

Coming from a job where everyone put on a show of strength, it was not easy for him to talk about how he was feeling, but eventually, with the help of friends and therapy, he began to find a way through the darkness:

> People think that PTSD is this Hollywood thing where you wake up with vivid dreams and shakes and sweats but it's not that at all. For a lot of veterans, myself included, it's not just flashbacks but more that feeling of belonging to something, that identity that you miss.

Knowing that he needed to feel as if he was doing something worthwhile, he signed up for another tour of duty, this time as a privately contracted bomb disposal expert in Syria.

Laidlaw described Syria as 'Afghanistan x 100 – absolute carnage'. He had never seen destruction on that level and it shocked him to the core. One day, through the cacophony of explosions and gunfire, he and his colleagues heard what sounded like a child crying. They rushed towards the rubble of a decimated school. Under a large concrete plinth they found the bodies of a dog and her puppies. They had all been killed. But where was the noise coming from?

Suddenly one of the men shouted, 'Dog! Dog!' There, amidst the rubble, dust and destruction, was a ball of fluff, a tiny Asian Shepherd cross puppy crying out for help. Laidlaw looked at it and said, 'That's Barry.' The name stuck, although the discovery that 'he' was actually a 'she' led them to 'girl it up a bit' to Barrie.

Laidlaw was immediately captivated. The dog looked so sad and lost that he was determined to give her the love and care she needed. She was hungry and thirsty so he shared his rations with her. After three days she grew to trust him, and he was able to pick her up and take her back to base. Barrie fell asleep in Laidlaw's arms.

The puppy needed to be walked and fed and trained, and soon everyone was eager to do their bit. The other men became Barrie's 'uncles'. Within the camp she helped create a very different world. Laidlaw says, 'I think within the first few days of Barrie being back in our camp, all of us realised how much having a puppy like this was helping.' He explains:

> She was a big distraction from everything we saw out there. There were weeks out there when you'd see 10–15

> dead bodies a day, men, women and children, so that in itself has its toll but just to come back to the office and have a play around, a mess around made everything better.

The dog would accompany Laidlaw on jobs in Raqqa, wearing a harness fashioned from a bulletproof vest. He was only too aware of how much Barrie's presence had helped him, and he was determined that he would take her back to Britain, no matter what:

> She had become this beacon of light at a time in my life when I was really struggling for a purpose, struggling to realise why I was here. I was Barrie's dad and I had something to look after and be responsible for – there was no chance she was staying in Syria without me.

Getting Barrie out of the country was not going to be straightforward. While sorting out the paperwork, Laidlaw flew to the UK to attend a wedding. Afterwards, as he headed for the airport to fly back to Syria, he got a phone call to say that the situation there had deteriorated. They were evacuating and he should stay at home.

He realised that he had just two weeks to get everything organised before the other men left, otherwise the dog would be stranded on the base all alone. They tried putting her on a supply truck to smuggle her over the border, but to no avail. He looked at whether the Americans could get her out and then he could pick her up from the US but that didn't work either. It was only with the help of a charity called War Paws, set up to help bring dogs home from war-torn areas, that Barrie finally made it to Iraq, then Jordan, where she spent three months in quarantine. Laidlaw kept the dog's photo as his screensaver and missed her so much that he would often find himself stroking his phone as if she were there beside him.

Finally she was flown to Paris where they could meet again. Barrie was traumatised after her journey and took a while to recognise that in front of her was the man who had rescued her in Syria and cared for her in her early days. When she realised that the person stroking her so lovingly was Laidlaw, she rolled on to her back for a tummy rub. Sean cried with relief and reflected on how this dog had changed his outlook: 'I was lost. I didn't know what was going on. It took a little puppy to ground me and make me whole again. She means the world to me and I couldn't imagine where my life would be without her.'

Laidlaw has no doubt that his newfound identity as Barrie's 'dad' has saved him from returning to the dark days in which PTSD threatened to destroy him:

> When I'm over-stressed and anxious, depressed, I don't just sit in my room staring at the ceiling thinking the world's caved in on me, I've got Barrie there jumping on me, trying to get me out for walks so I have to go out, I have to walk, I have to play with her. As annoying as it is at the time, an hour later I'm glad I've done that and got outside and she really helps drag me out of wherever I am.

It was a big adjustment for Barrie to learn how to be a pet rather than an animal of war, but like her owner, she has now managed to settle into her new life. Laidlaw describes her as 'my tiny saviour' and credits her for the fact he has now, at last, found happiness:

Meeting her was the best day of my life. Without her I don't know if I would have ever been able to climb out of that dark pit of despair after Afghanistan, to acknowledge the atrocities that I witnessed as a soldier or learn how to be a civilian . . .

Today, I work part time as an assistant paramedic and run a fitness training business with a friend. Although I still have moments when I can feel myself getting anxious, I just close my laptop and play with Barrie . . . With her around, I have clarity and a purpose. And although people say I saved Barrie's life, the truth is that she saved mine.

BARRY

It only takes a mention of the Swiss national dog, the St Bernard, for an iconic image instantly to spring to mind. A giant, deep-chested, hairy-coated, friendly-faced rescuer standing proud on a snowy peak, with a barrel of brandy attached to his collar. But is this an accurate picture of the dog famed for saving climbers and travellers from avalanches or treacherous Alpine passes? Not exactly.

To know more, we need to go back to 1800 and the Great St Bernard Hospice on the treacherous Great St Bernard Pass (there's a theme here) connecting Martigny in Switzerland with Aosta in Italy, 8,200 feet above sea level. The hospice, run by monks, was a haven, providing food and shelter to many a weary traveller for almost a thousand years.

Its dogs were gathered from the local farms and were

variously Alpine Mastiffs, Alpendogs, sacred dogs and hospice dogs. They received no designated training; the younger dogs would simply learn their search-and-rescue skills by following the more experienced ones. Between them they are said to have saved around 2,000 lives.

Severe winters meant that many of the hospice's dogs died in avalanches as they carried out rescues. This prompted the monks to begin a breeding programme in the mid-nineteenth century, and the dogs that remained were crossed with Newfoundlands. It was not until 1880 that the name St Bernard was used to describe the breed.

As a consequence, these early St Bernards looked very different from the ones we know today. They were smaller and much lighter – less than half the 290 pounds they might weigh today. That increase in weight has put them out of a job because they are too heavy to be lowered from helicopters so are no longer used to sniff out bodies in an avalanche. They are now more often to be found in care homes or schools, as they are wonderfully sociable dogs and can help people suffering from stress or anxiety.

As for the barrels, sadly they are the stuff of legend. An 1820 painting by Edwin Landseer, named *Alpine Mastiffs Reanimating a Distressed Traveller*, depicts two St Bernards trying to revive said victim, one wearing a barrel which Landseer claimed contained cognac. The image has stuck, and the barrel has become a logo almost, but no dog from the St Bernard Hospice ever wore one or carried any brandy to revive a frozen casualty – probably a good thing as alcohol and hypothermia are a bad mix. The dogs did carry milk from the milking shed for the monks of the St Bernard Hospice and that may have been where the image originated.

J. J. - 8927. - BARRY
Fidèle serviteur de l'Hospice du Grand St-Bernard

A contemporary postcard of Barry.

The most famous of all the St Bernard Hospice dogs, indeed the most celebrated rescue dog of all time, was Barry der Menschenretter (the 'people rescuer') who was born in 1800, and saved more than forty people in the course of his heroic career. His most famous rescue was that of a young boy whom he found unconscious and close to death in an ice cavern after an avalanche. After warming and rousing him by licking him, he managed to get the child onto his back and carried him to the warmth and safety of the hospice.

A taxidermist preserved Barry's body after his death in 1814 and he can still be found at Bern's Natural History Museum. In fact, he is quite the tourist attraction, particularly for travellers from Japan, who know all about him. Michael Keller, the vice-director of Bern Tourism, said:

> His story is still great, running around the mountains saving people. He's giving hope to everybody and I think

> it's this combination that makes Barry a good ambassador for Switzerland. It's like chocolate and cheese, I'd say Barry's in about third position and he's great for everybody.

Interestingly, the hallmark used for all precious metals and fineness standards in Switzerland is the head of a St Bernard – and while the symbol is not mandatory in law, the Office of Precious Metal Control files all refer to such hallmarks as 'Barry'.

The last Alpine rescue to involve a St Bernard took place in 1955, but the dogs' heroic rescues will never be forgotten. There is a monument to our hero in the Cimitière des Chiens (dog cemetery) near Paris, and the Barry Foundation has been set up to breed dogs from the hospice. There is always one named Barry in honour of the dog whose courage, kindness, determination and warmth set the standard for mountain rescue.

BOB

Bob is a modern, everyday hero of the animal world. His story is not one of military prowess or mountain rescue. He didn't hail from a stately home or aspire to high office. Bob never sought fame or fortune and yet both came his way. A string of best-selling books, a feature film based on his life and a reputation as a lifesaver, turned Bob from street cat to celebrity.

James Bowen, who wrote the books, also gives Bob the

credit for turning his life around: 'I've had cats all my life and I've never known one to be as intelligent as this one. He's extraordinary.'

Bowen had hit rock bottom. He had been sleeping rough, was addicted to heroin, and had been earning money by busking and selling copies of the *Big Issue*. He was trying to change things and was on a recovery programme, but life was tough when Bob found him.

James could barely look after himself: the last thing he needed was a pet. Bob, named after a character in *Twin Peaks*, first appeared curled up in the hallway outside his one-bedroom assisted-housing flat in North London in 2007. When Bowen fed the cat, he discovered Bob had a leg injury and fleas. He spent his last £30 taking Bob to the vet for treatment.

Once he'd recovered, James tried to send Bob on his way but the cat wasn't having any of it. After several attempts to follow James up the street, onto the bus and, during one hairy moment, across the street, James realised that Bob had claimed him. 'He needed me and I didn't know it then, but I definitely needed the love he gave me', James said.

Soon the two were inseparable and could often be spotted out busking together – Bob sitting on a rug as James sang and played. Buskers with dogs were common, but no one had seen one with a cat before. The pair became famous, with people uploading videos of them onto social media, which in turn drew more tourists to seek them out. It proved to be a turning point for Bowen, who was able to put his drugs and subsequent withdrawal treatment behind him. 'I believe it came down to this little man', Bowen says of Bob:

> He came and asked me for help, and he needed me more than I needed to abuse my own body. He is what I wake up for every day now . . . he's definitely given me the right direction to live my life.

The growing attention led to a story in the *Islington Tribune* in 2010. It attracted the attention of a literary agent who approached James to write a book. *A Street Cat Named Bob* sold over a million copies in the UK and was translated into thirty-five languages. Bowen has now written eight books which have become worldwide bestsellers. The first film, starring Bob as himself, won the award for Best British Film at the National Film Awards of 2017. Bob made many appearances with James in TV interviews and always took the limelight with his cool persona and his ability to give a high five on demand.

Bob was a cat unlike any other and his greatest trick was to transform the life of a man who didn't think he'd get a second chance. The two of them lived happily together in Surrey where James built a special 'catio' for Bob to bask in the sunshine. Bowen has continued to write, and together he and Bob raised huge amounts of money for homeless charities and animal welfare organisations.

Bob the cat died on 15 June 2020 at an estimated age of fourteen. James said:

> Bob saved my life. It's as simple as that. He gave me so much more than companionship. With him at my side, I found a direction and purpose that I'd been missing. The success we achieved together through our books and films was miraculous. He met thousands of people, touched millions of lives. There's never been a cat like him. And never will be again.

I feel like the light has gone out in my life. I will never forget him.

BOBBIE THE WONDER DOG

One of my favourite books as a child was *The Incredible Journey* by Sheila Burnford, about a group of three animals (two dogs and a cat) who embark on a journey across the Canadian wilderness. They overcome all sorts of dangerous foes and situations as they travel over 300 miles to their human family and their home. I always thought it was amazing that animals seem to know the way home and to navigate over vast distances. Whether it's beetles or birds, turtles or termites, they don't need a map or a satnav to tell them the way.

In April 2016 Pero, an adventurous four-year-old working sheepdog, decided he didn't want to stay on the farm in Cockermouth, Cumbria, where he had been sent to help with the sheep. Instead, he made his way over 240 miles back to the place where he was born near Aberystwyth, on the coast of mid Wales. It took him two weeks and he was beside himself with joy to meet up with his original owner/breeder, Alan James. A microchip check confirmed that young Pero really had made it back to his first home.

This ability to find home comes down to 'personal bonds' between an owner and their dog and to the dog's own biology. Their olfactory ability can do more than sniff out buried bones, fugitives or drugs, it can also give a dog an awareness of its location. Dogs are also very reward driven – and their positive associations with particular places or people are likely to result in the overwhelming desire to return to them.

Some dogs take this homing ability to an exceptional degree.

In the early 1920s, America was enthralled by the story of Bobbie the Wonder Dog. Bobbie was a two-year-old Collie cross, the beloved pet of the Brazier family. In the summer of 1923, the family packed up their Overland Red Bird touring car and headed east to Indiana on holiday. And, of course, Bobbie went with them – perched proudly atop the pile of luggage on the back seat. They were at a petrol station when a pack of local mongrels jumped on Bobbie.

The last Frank Brazier saw of his beloved dog, was Bobbie running for his life with three snarling hounds in hot pursuit. The Braziers looked everywhere for Bobbie. They called around town, advertised in the local newspaper and drove all over in search of him. No one had seen him. There were no clues. He had disappeared. Eventually, they regretfully left, leaving instructions that should he turn up, they would pay for him to be sent home by rail.

Desolate, the family returned to Oregon. They had given up all hope when in February 1924 – six months after they had left Indiana – a bedraggled Bobbie turned up on the doorstep of their diner, the Reo Lunch Restaurant.

It was amazing he had survived his trek. He had walked over 2,500 miles across eight states, crossing mountains, desert and prairie. He had swum through rivers and crossed the Continental Divide in the dead of winter. Not surprisingly, he was in very poor condition: weak, dirty and emaciated, his paws worn right down. The Braziers were dumbfounded and delighted.

Newspaper cutting celebrating Bobbie's return.

Within a week the story was making national headlines. Friendly people with whom Bobbie had stayed for a night or two on his journey wrote in to tell their stories of how this strange dog had appeared and then disappeared, seemingly on a mission.

With all this information, the Humane Society of Portland was able to piece together a surprisingly precise account of the route Bobbie took. He had first followed the Braziers northeast, farther into Indiana. Then he started striking out on what must have been exploratory journeys in various directions – perhaps trying to pick up a familiar scent to give him a sense of the direction to take. Eventually, he found what he was looking for and made for the West Coast. On their trip, the Braziers had left their car in service stations every night. Bobbie visited each of these on the way, along with a number of private homes. In Portland, he stayed for some time with an Irish woman who found him with legs and paws in a terrible state and nursed him back to health.

As the story of Bobbie spread around the United States, fan mail began to arrive from admirers all over. Now nicknamed 'Bobbie the Wonder Dog', he was featured in books and articles and he even became a film star. Gifts flooded in, among them a jewel-studded collar. Silverton gave him the key to the city, along with special permission to walk its streets free from fear of the municipal dogcatcher.

When Bobbie died in 1927, he was buried with full honours. More than 200 people attended his funeral and a wreath was laid on his grave by the famous canine film star, Rin Tin Tin the German Shepherd.

BUCEPHALUS

When we were very young, my brother and I shared a joke every morning at breakfast. My absolute favourite, which made us both cry with laughter, was as follows:

> QUESTION: What's round and green and conquered the world?
>
> ANSWER: Alexander the Grape.

It still makes me chuckle now and was part of the reason I became interested in Alexander the Great. Whatever gets you going, eh?

Alexander the Great was history's greatest general and famously never lost a battle. By the age of thirty, he had created one of the largest empires of the ancient world, stretching from Greece to northwest India and from Macedonia to Egypt. Even more fascinating to a young equestrian like me, his horse Bucephalus was believed to be the key to his success.

According to the historian Plutarch, in 344 BC, when he was around twelve or thirteen years of age, Alexander won the horse in a bet with his father, King Philip II. Bucephalus, valued at three times the normal price of a warhorse, had been brought to Macedonia by Philoneicus of Thessaly. A huge black stallion with a gleaming coat, he was impressive but impossible to manage, rearing up whenever anyone came near. King Philip ordered him to be taken away.

The horse objected, his eyes popping out of his head in fear and rage as the king's attendants struggled with him. Alexander rose from his seat and called on them to stop. He questioned why such a fine horse should be rejected just because of human weakness. He challenged his father, saying that he would pay the thirteen talents for the horse himself, if he proved unable to tame him.

In the sunshine, Alexander could see that the horse was scared of the large black shape that seemed to be following him – his shadow. He turned Bucephalus towards the sun, so that his shadow fell behind him, and hopped onto his back. The mocking laughter of the onlookers changed to cheers as Alexander took the reins and rode away. The historian Plutarch maintains that this was the turning point in his life – the birth of Alexander 'the Great'. When Alexander returned to the arena and dismounted, Philip said: 'Oh my son, look out for a kingdom equal to and worthy of thyself, for Macedonia is too little for thee.'

So began a lifetime of exploration, conquest and empire building. Bucephalus and Alexander became inseparable. The horse would allow no one else near him and Alexander wanted no other steed in battle. Together, they conquered Asia.

On a rare moment of separation after they had defeated Darius III of Persia, Bucephalus was kidnapped. Alexander went ballistic and threatened to slaughter every person, fell every tree, ravage the land and lay the country to waste. The kidnapper had second thoughts and rapidly returned the horse, begging for mercy.

When Bucephalus died of old age after the Battle of the Hydaspes River in 326 BC, Alexander mourned his death deeply and founded a city at the foot of the Himalayas in his memory called Bucephala, or Alexandria Bucephalous.

Such was the cult of Alexander the Great and Bucephalus that it became fashionable for other leaders to single out a favoured horse. Julius Caesar's was Genitor (named in honour of his father) while the Roman Emperor Caligula was so fond of his horse Incitatus, that he held birthday parties for him, bought him gifts including an ivory manger and a house, and planned to make him consul.

Bucephalus, meanwhile, has not been forgotten and features in many a book, song and film. His image graces a Charles Le Brun painting hanging in the Louvre, and the bicycle in the TV series *Father Brown* also bears his name.

CAIRO

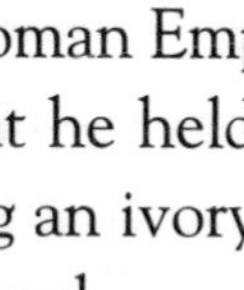

Until recently, the story of this canine hero was something of a mystery. Not, perhaps, surprising when you learn that he played a crucial part in one of the most intense military operations in modern history, Operation Neptune Spear – the plan to assassinate the founder and first leader of the terrorist group Al-Qaeda.

For ten years after the attacks on the Twin Towers in New York known as 9/11, the Americans had been searching for Osama bin Laden. Cairo, a Belgian Malinois, was the only four-legged member of the US Navy SEAL Team Six mission that stormed his secret compound in Pakistan on 2 May 2011.

While much of the detail of SEAL dog training is shrouded in secrecy, for obvious reasons, it's understood that they work alongside human members of the team, learning to parachute (solo if landing in water), to detect people hiding inside buildings and to check for explosives or booby traps with impressive speed. They perform a crucial role and are so highly valued that in 2018 General David H. Petraeus, commander of the US forces in Afghanistan, asserted that 'the capability they bring to the fight cannot be replicated by man or machine'.

There are around 600 dogs serving in the US military in Afghanistan and Iraq, a number that's expected to keep on growing. They wear protective body armour and heat-vision goggles, allowing them to see body heat through concrete and brick. They sometimes sport special vests, equipped with infrared or night-vision cameras, so that handlers can monitor exactly what they are seeing from as far as 1,000 yards away. They offer other strengths too. Belgian Malinois and German Shepherds are able to run twice as fast as their human counterparts – great for catching fleeing enemy forces – and are trained to hold on to suspects by biting or overpowering them, depending on their size. Other canine counterparts, including Labrador retrievers, head out in front of patrols to check the safety of the route. It's a dangerous life.

Cairo was an unlikely and unusual hero. He and his handler, Will Chesney, almost didn't end up as a pair at all. Cairo was slightly stand-offish and Chesney preferred another dog called Bronco. However, the programme director had spotted that Cairo had special potential. The pair eventually bonded during a seven-week training camp in California. In fact, as early as the second night, Cairo decided to share his new master's bed, although Chesney was rather less keen about this, describing the dog as 'a blanket hog'.

The pair's main job was to locate hidden enemies. Now retired, Chesney can finally shed a little more light on how his canine compatriot saved his life on a number of occasions and how his skills were superior to those of any other dog he had come across. Cairo was alert, instinctive and fearless. He would always react first. 'Over and over again, there'd be a guy waiting a couple of feet away to pop out, and the dog would find them', Chesney said.

In June 2009, Cairo was seriously injured. The pair had been seconded to Afghanistan and were taking part in a battle against heavily armed insurgents who fled to the shelter of a wooded ridge as the dog picked up their scent and set off in pursuit. As he heard another round of gunfire, Chesney called to his dog to come back, but Cairo was nowhere to be seen. Panicked, Chesney kept on calling, and eventually spotted the dog emerging in the distance, then making his way slowly back to his owner before collapsing at his feet.

Cairo, who had been shot in the chest and front leg, was bleeding heavily and struggling to breathe. Chesney made the call on the radio 'FWIA' (friendly wounded in action). It didn't matter that Cairo was a dog, his status was the same as a human SEAL. He was taken back to base by helicopter, where doctors and nurses (rather than vets) fought to save his life. It was touch and go, but Cairo's strength and determination helped him to make a full recovery, and to return to front-line duties with his handler.

The bin Laden siege was the riskiest mission of their entire career: all the SEALs involved were advised to get their affairs in order as they trained at a full-scale model of bin Laden's compound in North Carolina. But both Chesney and Cairo lived to tell (at least part of) the tale.

Two helicopters carried the SEALs to Abbottabad, the dog dressed for action in a Kevlar vest and night-vision goggles. One landed safely, allowing the men to invade the compound while Cairo and Chesney formed part of the team which would search the perimeter for bombs or escape tunnels before entering the house. The other Black Hawk crashed inside the compound and was badly damaged, though everyone on board escaped without injury.

Cairo and Chesney searched the first two floors before one of their comrades gave them the news that bin Laden was dead. The operation should have been quick and quiet but the helicopter crash had woken most of the neighbourhood and alerted people to what was going on. It was Cairo who kept them at bay as the other SEALs seized papers, photographs and computers before blowing up the helicopter and getting out of there as quickly as they could.

Back in the US less than two days later, the men were meeting President Obama, who up till then had not even been aware of Cairo's existence, let alone his critical role in the successful mission. The Secret Service had ordered that he be muzzled in another room. When the President heard about Cairo, he said, 'I want to meet this dog.'

Chesney is still frustrated that the role of dogs in modern-day warfare goes unmarked. Every SEAL who took part in the bin Laden raid was decorated with a Silver Star – except Cairo. His owner told the *New York Post*, 'It was a disappointment to me. He was every bit as important to the mission as anyone else. He risked just as much.'

When Chesney left the military in 2011 he missed Cairo dreadfully. He was suffering from Post Traumatic Stress Disorder, often anxious and erratic in his behaviour. The only thing that helped was seeing Cairo. He visited him regularly and made it clear he wanted to adopt him but, as always when bureaucracy takes over, it wasn't that straightforward. Chesney heard that two other SEALs had expressed an interest in having Cairo. He panicked and even thought up ways he could kidnap him.

It took a year of paperwork and box ticking but eventually, in 2014, Chesney and Cairo were reunited. Chesney believes that even out of service, the dog saved his life by getting him through PTSD, anxiety and depression. Sadly, their life together didn't last long. Cairo had aged before his time and had been through a lot. In April 2015, Chesney held his partner's paw as he was put down. His ashes are in an urn with his pawprint on it and Chesney still has Cairo's bloodstained harness as a memento of his brave and loyal comrade.

CANARIES

Small but mighty, these noisy, bright little birds have helped save thousands of human lives. Originally from the Canary Islands, the male birds have a sweet song and were bred in captivity in the seventeenth century to be sold as exotic pets. They were particularly popular amongst the aristocracy in Spain and, from there, Spanish sailors brought them over to the UK.

If they amused the upper classes in the eighteenth and nineteenth centuries, by the early twentieth they had become indispensable to members of the labouring class, those who worked down the mines.

Our story starts with John Scott Haldane, a Scottish physiologist, 'the father of oxygen therapy', notorious for the risky experiments he carried out on himself and his son. These included locking himself into a sealed chamber, inhaling various (potentially lethal) cocktails of gases, and recording the effects they had on body and mind. He devised both the decompression chamber that allowed deep-sea diving and the oxygen tent. After being sent to the front by Lord Kitchener in the First World War to investigate the poison gases being used by the Germans, he invented early forms of gas masks and the first respirator.

He was particularly interested in coalmining, which was a hazardous activity in the nineteenth and early twentieth centuries. Thousands of labourers risked their lives every day, going deep underground to mine the coal. Disasters caused by explosions, tunnels collapsing or exposure to dangerous gases happened frequently. There were hundreds of deaths every year. Haldane investigated numerous disasters, and after examining the bodies of miners killed in pit explosions, he identified their killer as carbon monoxide.

He then tested the effect of the gas on his own body in a closed chamber, vividly describing the results of his slow poisoning. In the nineteenth century, miners relied on rudimentary sensors such as oil lanterns to alert them to the presence of dangerous gases but there was no test for carbon monoxide. Haldane suggested the use of canaries as a sentinel species. He suggested that the birds would be perfect for spotting the first signs of toxic gases underground because they needed such immense quantities of oxygen to enable them to fly – and fly to heights that would cause altitude sickness – their anatomy allowing them to get one dose of oxygen when they inhale and another when they exhale, by holding air in extra sacs.

A rescue miner with a canary in Wales in 1982.

Small birds were very popular pets at the time and the miners enthusiastically welcomed the idea. They were much less keen on Haldane's other suggestion of white mice, which were too much like the dreaded rats that plagued the mines. Many of the rescue stations set up in the first decade of the twentieth century soon contained aviaries. By 1914, mining authorities proudly boasted that 'Canaries save about 800 lives a year.'

The miners valued their colourful noisy pets and kept a close eye on them. As soon as they stopped singing or showed any sign of distress, the miners rushed to help them. The birds were the first to receive treatment. Some were even kept in special carriers with small oxygen bottles attached to revive them. This early-warning practice, subsequently adopted by other nations including Canada and the USA, was in use until relatively recently. The miners were very reluctant to lose their tuneful colleagues and there were still 200 working canaries in British mines as late as 1986 when the law finally changed. Today they use hand-held electronic sensors instead.

In a non-mining incident, another canary can be honoured here for saving lives. The bird doesn't have a name in the small entry in the *New York Times* of Saturday, 7 April 1906 but what is clear is that this little bird saved a whole family from certain death.

John Bietze, his wife and their young daughter Ida lived in Middletown, NY, and allowed their pet canary to fly freely around the house. When his shrill chirping from the end of the bed woke Bietze from an unnaturally heavy sleep in the early hours, he discovered that gas from their downstairs stove was escaping and had already filled the rooms where the family slept. The canary had woken him in the nick of time. With enormous effort, he managed to rouse his wife, and then found their daughter, unconscious and gasping for breath. He dragged them both to the clean air outside and to safety. When he went back to rescue their canary, the heroic bird had received a fatal dose of the gas and, sadly, was dead.

CHARISMA

Sometimes you can see a horse's name and know that it's going to be a champion. Charisma certainly had that ring to it, but his nicknames did not. 'Podge' and 'Stroppy' were how he was known at home and his rider Mark Todd described him as a 'fat, hairy pony'. Those words were said with love, because put the little fatty into competitive mode and he became Charisma the class act. He was, arguably, the greatest eventer of all time.

Charisma was born in October 1972, to a show-jumping, polo-playing mum called Planet who had the rare distinction of being the first mare in New Zealand to jump her own height. Small wonder, then, that four-foot fences were not enough to keep her young son securely in his paddock. Charisma was jumping for fun rather than freedom. He did it just because he could. Todd was working as a farm hand when he saw the youngster and thought what a shame it was that he wouldn't be big enough to compete.

Charisma's career began in Pony Club competitions, then on to more prestigious competitions in eventing, show jumping and pure dressage. He wasn't big at fifteen hands, three inches, but he was versatile.

His magical partnership with Todd started by accident and quite late on. Charisma was ten years old by the time they hooked up in May 1983, when Todd's horse was sick and he was offered the chance to ride Charisma instead. It was not the most likely combination with Todd a towering six foot three and Charisma barely bigger than a pony. What the little horse lacked in height, he made up for in girth.

Todd's first impression of Charisma as a fully-grown gelding was unpromising. He described him as a 'plain-looking little horse'. However, when he had the chance to ride him, he swiftly changed his opinion because he 'moved beautifully'. Charisma could pile on the kilos if he wasn't in full exercise, and an equine crash diet was required. Even his bedding needed to be changed to newspaper strips so that he didn't help himself to a midnight feast of straw.

The work put in – on both sides – was more than worth it. The pair won their first two one-day events and took gold at both New Zealand's One Day Event Championship and the National Three Day Event. That was enough to secure them a spot on the team for the Olympic Games, but first there were challenges Todd wanted to take on in the UK. He wanted to ride Charisma at Badminton, the biggest and most famous horse trials in the world.

On the journey over, Charisma was ill and developed a sinus problem from which he never completely recovered. He still gave his all and managed to finish second at Badminton behind Lucinda Green on Beagle Bay. On to Los Angeles for the 1984 Olympics where after a decent dressage test, Charisma fairly charged round the cross-country course. 'Podge pulled like anything', Todd wrote in his book *Second Chance: The Autobiography*:

> He would tuck his head in like a bull and charge, and there was little one could do except pray he wouldn't make a mistake . . . Unsurprisingly, he had the fastest cross-country round of the day and, despite the boiling temperatures, finished easily inside the time.

A clear round in the show jumping and a rail down from the leader secured Todd and New Zealand their first equestrian Olympic gold medal.

Before the 1984 Olympics, Charisma's owner Fran Clark had been trying to sell the horse and, after a falling out with Todd, wanted to do the deal with anyone but him. After they had won gold together, this became a very public scandal and the eventing world rallied round Todd, refusing to take his horse from under his legs. His great friend Lizzie Purbrick helped out by pretending she was buying Charisma. Todd's sponsor transferred £50,000 into her account, she sent it on to Clark and the horse was sold. For Todd, it was a huge relief. The idea of losing the ride on Charisma had been unbearable.

In 1985 they had three good one-day-event wins before finishing second again at Badminton, this time to Ginny Holgate on Priceless. Charisma had led after the cross country but one rail down in the show jumping denied him the prize. In 1986 they travelled to Australia for the World Championships in Gawler but had a rare fall at the water jump. Later that year, they won the big three-day event at Luhmühlen in Germany.

As the clock ticked on to the next Olympic Games, Todd's major concern was that at sixteen Charisma might be too old for a long journey and a taxing competition in a warm climate. The horse, however, didn't know how old he was and his form in 1988 was better than ever. He did a brilliant dressage test in Seoul, went clear inside the time cross-country and had two fences in hand for the final show-jumping phase. That was plenty and Charisma became the first horse to win back-to-back gold medals since 1928. He was retired at the end of that year and went home to New Zealand to embark on a gold-medal tour. The pair had become national heroes.

Even though the Commonwealth Games does not include equestrian events, Todd and Charisma were given a major role at the 1990 Games in Auckland, carrying the Queen's baton into the stadium at the opening ceremony.

The bond between horse and rider was legendary, and Charisma would follow his owner around like a faithful puppy. Charisma lived to the age of thirty on Todd's farm. Todd wrote of him:

> My horse was Charisma both by name and nature and I adored him. We were such an unlikely combination – he was cute and small and I was tall and lanky. It was such a good story – a little horse just doing Pony Club in New Zealand travels across the world and wins the Olympics – that we captured everyone's imagination.

CHER AMI

The role of pigeons in war goes back many centuries. Cyrus the Great used them to communicate with parts of his Persian empire in the sixth century BC. It's said that Julius Caesar employed the birds to send news during his conquest of Gaul. In the early thirteenth century, Genghis Khan, Supreme King of All the Mongols, used pigeon post to send news of his military campaigns across Asia and Eastern Europe. In 1815, a carrier pigeon owned by Nathan Rothschild is said to have carried home news of Napoleon's defeat at the Battle of

Waterloo (allowing him to invest in British government bonds before any other traders could get a look in).

Pigeons were lifesavers during the Siege of Paris, when French citizens were encircled by the Prussian Army for four and a half months during the freezing winter of 1870–1. Cut off from the rest of the world, life in Paris grew increasingly dire, its famished citizens forced into eating pets and vermin. Communication with the rest of the country became ever more essential. When a gas-powered balloon was successfully used to carry mail from the capital to the outside, three homing pigeons were sent out with a second balloon, the Ville de Florence. Thus a two-way messaging system was established.

More than 400 pigeons were sent out in the course of the campaign, of which just seventy-three returned. As tribute to the birds and their balloon pilots, a metal statue, designed by Frédéric Bartholdi, the sculptor of the Statue of Liberty, was erected in January 1906 at the Porte des Ternes. Sadly, it was melted down for munitions by occupying German forces during the Second World War. In his book *Inside Paris During the Siege* (1871), Denis Arthur Bingham described the pigeons' 'heroic defence' of Paris being 'the admiration of the whole civilised world', adding 'the birds have become sacred to all patriots'.

By the beginning of the First World War, the role of pigeons in military manoeuvres was firmly established on both sides. The average dog can only run between two and five miles, but a good racing pigeon will fly more than sixty miles to get home. Crucially, they can fly *over* not through the enemy lines. Pigeons were vital to the war effort.

Between 1914 and 1918 the French Army had called up more than 30,000 racing pigeons, with the threat of a death sentence hanging over anyone impeding their flight. One called Le Vaillant, a member of the US Signal Corps, was awarded the Ordre de la Nation for saving the lives of 194 men during the Battle of Verdun.

Another American army bird, Cher Ami, a Black Check cock, was one of 600 birds used to carry messages and conduct surveillance by the Americans during the First World War. He carried his messages in a tiny metal tube attached to his leg, each message kept as short as possible to keep the weight down. Originally donated by British pigeon fanciers, he carried twelve crucial messages in total, the most important of which led to the salvation of the surviving members of Major Charles W. Whittlesey's 'Lost Battalion' of the 77th Infantry Division.

The battalion were heavily outnumbered by the Germans and couldn't shoot their way out because they had run out of ammunition. As the day wore on the casualties increased and the soldiers realised that they were being shelled not only by the enemy but by their own side who had failed to recognise them. By the next morning they were desperate and sent for help.

Their first message read, 'Many wounded. we cannot escape', but the bird was shot down as soon as it took off. A second message, 'Men are suffering. Can support not be sent?' had the same fate. Cher Ami was the battalion's final pigeon and last hope of survival. The crucial missive – reading, 'We are along the road parallel to 276.4. Our own artillery is dropping a barrage directly on us. For heaven's sake stop it.' – was attached to his leg and he took off. The sky was full of bullets and he took fright, landing in a tree nearby. One of the soldiers had to shimmy up the tree to persuade him to fly back to his dovecote. Eventually, he got the idea and set off.

Cher Ami on one leg was awarded the Croix de Guerre.

The enemy focussed a hail of bullets at him. He took a bullet to the chest and fell to the ground but then miraculously took off again. He managed to fly twenty-five miles and safely delivered the message, dangling from his broken leg. A rescue mission was immediately launched to bring back the Lost Battalion and army medics worked on Cher Ami's leg and his eye, which had been blinded in the attack.

Thanks to the bravery of Cher Ami, 194 men safely made it back to American lines. The pigeon was awarded the Croix de Guerre for his heroic dedication to duty and returned to the US where he was fitted with a wooden leg. When he died, eight months later in 1919, his body was preserved and is now on display at the Smithsonian Museum of American History in Washington DC.

In 1931, Cher Ami was inducted into the Racing Pigeon Hall of Fame and he was also awarded a gold medal by the Organized Bodies of American Pigeon Fanciers in recognition of his war service. In 2019 he was honoured at a ceremony on Capitol Hill, Washington DC, as one of the first recipients of the Animals in War & Peace Medals of Bravery.

There is to this tale, a twist. When Cher Ami's body was mounted by the taxidermist, it was discovered that she had been *Chère Amie* all along.

CHIPS

As with pigeons, the use of dogs in war has quite a history. The Ancient Egyptians, Greeks and Romans all used dogs in one way or another, whether to patrol, act as sentries or even do their bit in battle.

The earliest recorded use of canines in combat goes back to 600 BC, when Alyattes of Lydia, an Iron Age kingdom situated in what is now Turkey, used dogs to ward off invaders. Fast forward to 434, when Attila the Hun favoured ferocious Molossers, dogs similar to mastiffs, for his campaigns.

The names of the majority of the dogs who fought through the ages have been lost to war, though, in the early nineteenth century, Mous (short for Moustache), a black Barbet from Normandy, became famous for his part in the French Revolutionary and Napoleonic Wars. Mous latched on to a regiment of French Grenadiers and followed them through the Italian campaign, successfully warning them of attacks. Poor thing, he lost an ear at the Battle of Marengo, a leg at the Battle of Austerlitz, where he retrieved the regimental standard, and was eventually killed by a cannonball at the Battle of Badajoz in Spain.

More recently, dogs have been revered and celebrated for their heroic contributions to war. The most decorated dog of the Second World War was Chips, one of only eight recipients of the Animals in War & Peace Medal of Bravery. A German Shepherd/Collie/Siberian Husky cross, Chips was born in 1940 and was donated to the US Army war effort by his owner, Edward J. Wren of New York, one of many who offered up their family pets to serve their country. 'It killed my mother to part with him', admitted Wren's son John, a toddler at the time. 'But Chips was strong and smart and we knew he'd be good at Army duty.'

Chips is rewarded with a doughnut, c.1944.

The dog was one of 10,425 to serve in the US Army's new 'K-9 Corps' and in 1942 was sent to the War Dog Training Center in Front Royal, Virginia, where he was drilled in the skills needed to be a sentry. Chips served alongside the 3rd Infantry Division in North Africa, Italy, France and Germany. He was the guard dog for the Casablanca Conference between President Roosevelt and Winston Churchill where the two men mapped out Allied strategy. His most dangerous mission came late in 1943, during the Operation Husky invasion of Sicily. His platoon, upon landing on the beach, were fired upon. Chips broke away from his handler, Private John P. Rowell, and made for the thick concrete pillbox where the enemy gunners were housed. Somehow he managed to get inside the pillbox and there he attacked the four Italian soldiers, forcing them to leave their cover and surrender to US troops.

Chips sustained powder burns and a scalp wound in the skirmish, but this didn't stop him returning to duty immediately afterwards. Later that night he heard enemy soldiers approaching and woke his men who captured ten further Italians prisoner thanks to his warning.

His actions brought him to the attention of General Eisenhower who, on meeting Chips and his keeper, said to Rowell, 'I hear your dog has performed heroically', and reached forward to pat the dog's head. Chips, suspicious of the stranger who was making his handler nervous, bit him. He wasn't to know that Eisenhower was on his side.

Chips's bravery earned him the Distinguished Service Cross (the US Army's second highest military award, given for extreme gallantry and risk of life in combat with an armed enemy force), a Silver Star (the third highest decoration for valour in combat) and a Purple Heart. After the war, a huge controversy erupted over giving military service awards to animals and the medals were revoked. However, his unit did their best to make up for this by unofficially awarding him a theatre ribbon (medal) with an arrowhead to denote an assault landing and battle (or service) stars for each of his eight campaigns.

In December 1945 Chips was officially discharged and returned to the Wren family, where he lived happily until his death a year later, although Wren said he no longer wagged his tail as much as he had done before he went to war.

In 2018 he was posthumously awarded the PDSA Dickin Medal at a ceremony at the Churchill War Rooms in London attended by John Wren.

CHRISTIAN

At the tail end of the swinging sixties, London was buzzing in psychedelic glory. Gone was the gloom of the post-war years, the economy was booming, the place was experimental, eccentric and exotic. It was also bursting with youth: 40 per cent of the population was under twenty-five, full of hope and dope. Dubbed by *Time* magazine the 'swinging city', London was the undisputed capital of style.

The hotspots of Carnaby Street and the King's Road were filled with packed boutiques and Harrods in Knightsbridge had the reputation of being the place where you could buy anything. And it was. From pianos to children's shoes, lunch and diamond rings. Their famous pet emporium had opened in 1917 and had already made its mark. Before the 1976 Endangered Species Act, it sold all sorts of wild animals, including panthers and tigers. In 1951 Noël Coward had an alligator bought for him there as a Christmas present and in 1967, future US President Ronald Reagan rang up to buy a baby elephant named Gertie.

In 1969, two young Australians saw a lion cub in a tiny cage at Harrods and knew they had to get him out of there. They paid 250 guineas (which is now around £3,500) and walked out of the shop with the lion on a lead. They called him Christian (because of Christians being fed to the lions) and took him to live with them in a flat above a furniture shop in Chelsea. It sounds crazy but it's true.

John Rendall and Anthony 'Ace' Bourke had been friends since childhood and met up when both came to London looking for adventure. Both animal lovers, they had not planned on the adventure being centred on a lion, but Christian changed their lives.

After the pair persuaded the bosses at the furniture shop downstairs – where they both worked – that it would be fine to have a lion living in the flat above, Christian moved in. He settled in remarkably quickly, sitting on their laps as they tried to read the paper and learning to use an extra-large litter tray within days.

All too soon, Christian grew a little large for their home, so Rendall and Bourke moved him to new living quarters in the basement of the furniture shop, the aptly named Sophisticat. The playful lion made the shop staff his unofficial pride, something the owners and manageress were more than happy to go along with. The cleaning lady was less keen after Christian started to steal her vacuum cleaner and dusters.

At night, Christian would sit in the shop window and watch the cars go by. He became a favourite with local children, who would flock to see 'their' lion whenever they could.

He went to restaurants and parties, hitching a ride in the back of Ace and John's convertible Mercedes.

A local vicar gave permission for him to be exercised in the graveyard of the Moravian church in the World's End part of the King's Road. The arrangement was short-lived, ending the day Christian jumped on top of the vicar's car and refused to move.

When the weather was nice, the three would head off to the seaside for a day trip. 'He was beautifully behaved', Rendall told the *Guardian* in a 2011 interview:

> And though he never bit or hurt anyone, you underestimated his strength at your peril. I remember taking him to a party once and he jumped on a friend he hadn't seen in a while and when he put his paws on her shoulders, one of them slipped, his claw got caught in the straps of her dress and the whole thing was on the floor.

The pair knew that their pet could not be with them for ever. They had always planned to keep him for a maximum of twelve months, before finding a more suitable home for the king of the jungle. They looked at Longleat, newly opened as the first safari park outside Africa, and considered sending him to live on a country estate. However, they were presented with the perfect solution when Bill Travers and Virginia McKenna, the stars of *Born Free*, came into the shop to buy a desk. Travers and McKenna had had no idea that Sophisticat housed a lion as well as a selection of pine furniture, and were not hugely impressed to find a wild animal living in the centre of swinging London. They knew exactly the right person to help Christian live a life in the wild.

Enter conservationist George Adamson, and his wife Joy,

who were famed for raising Elsa the lioness in the late 1950s, then successfully releasing her into the Meru National Park. Adamson, who became known as Baba ya Simba (Swahili for Father of the Lions), had retired as a wildlife warden in 1961. He now devoted his time to raising and rehabilitating captive or orphaned lions before returning them to their natural habitat. He moved to the remote Kora Nature Reserve in northern Kenya in 1970, and it was there that Rendall and Bourke took Christian to begin his new life.

For the first few days Christian was like a teenage tourist, hanging out on Rendall or Bourke's camp bed. But the timing was perfect. Boy, a lion used in the filming of *Born Free*, had been injured and was recuperating at Kora. This would offer Christian the opportunity to learn the skills he would need to survive, from a mature lion who had already experienced life out in the bush. The Australians were sad to leave their beloved pet but knew that they had brought him to the right place. Rendall said:

> To see Christian in his right environment was so exciting. Suddenly, instead of being 'exotic' he just fitted in, blending into the landscape. Even so, it was wrenching to leave him behind knowing all the inevitable dangers and hardships facing an animal in the wild, particularly a pampered one.

Adamson discovered that his new charge instinctively knew how to stalk and get thorns out of his paws, but he still had a lot to learn. The first step in Christian's integration into the normal life of a lion was to put together the nucleus of a new pride, which Adamson did by introducing him to an older male, Boy, and then to a female cub, Katania.

Things did not go to plan. First, Katania was swept away as they crossed a river and eaten by crocodiles. Then Boy was shot by Adamson after fatally wounding an assistant. The lion was severely injured, putting paid to his ability to socialise with man or beast. Adamson was supposed to be safely rehabilitating lions, but instead people were being killed. His reputation was on the line. Eventually, he found two new lionesses and was able to re-establish the pride, with Christian as its new head.

A year later, Rendall and Bourke travelled to Kenya to see how Christian was adapting to life in the wild. They had been warned that it would probably be a wasted trip as there was little chance he would react to them. The home-movie footage of the ensuing reunion is one of the most heart-warming things you will ever see.

The lion makes his way slowly down the rocks and then it dawns on him. Those aren't just any old guys standing watching him – they're *his* boys. He runs towards them and hugs them, moving from one to the other with his giant paws on their shoulders, rubbing the side of his head on theirs. He is ecstatic. The footage was uploaded to YouTube in 2008 and has been seen since by over 100 million people. It's a tear-jerking reunion between those who have loved each other and still do, with no fear or hesitation, and it's a testament to the way in which Rendall and Bourke raised the cub.

Rendall and Bourke published a book in 1971: *A Lion Called Christian*. They went to Kora one last time in June 1972. Adamson hadn't seen Christian for three months and was unsure he would appear. He warned the men again that it was unlikely, now that he was three years old and had cubs of his own, that Christian would recognise them. Within a couple of days of their arrival, the lion ambled into camp.

He was much bigger, full, mature and supposedly more dignified but Rendall said, 'He knocked George over, jumped on the table and interrupted dinner. He tried to sit on our laps, even though he was now a 500-pound cat.' All night long he stayed with his former owners, playing with them and rolling around in joy. A few days later, he left to rejoin his pride and they never saw him again.

When Christian was released into the wild there were estimated to be around 250,000 lions in Africa. Now there are less than 20,000. It is a species that must be protected, or it will become extinct within our lifetimes.

CLARA

Rhinos and haute couture are not natural bedfellows. Then again, they do say that there's an exception to every rule, and in this case the rule-breaking rhinoceros was a female Indian named Clara.

Born in Assam in 1738, Clara was less than a week old when her mother was killed by hunters. She was adopted by Jan Albert Sichterman, a director of the Dutch East India Company, and allowed free rein of his family's estate, including the house. As she was used to humans and had been hand reared, she was quite keen on being around them and would delight dinner-party guests by joining them at the table, eating from a porcelain plate.

Inevitably, small rhinos grow into great big rhinos and when Clara got too large to be indoors without creating havoc, Sichterman passed her on to a colleague, Douwe Mout van der Meer, a Dutch sea captain who sensed a business opportunity back in Europe.

Surviving the seven-month voyage to Rotterdam was no mean feat. Rhinos rely on mud to keep their skins from dehydrating – not something you can easily access while on the high seas. Seawater offered no substitute, and drinking water was scarce, way too precious to pour on a pachyderm. The sailors decided to seek an alternative lubricant and slathered her in fish oil, of which they had a plentiful supply. This made her a tad whiffy, especially during the warmer parts of the journey, but seemed to be effective.

Throughout the voyage, Clara lived in a cage on deck. Rhinos are generally solitary animals. They can run at speeds of up to thirty miles an hour and are used to defending themselves from potential predators with their horns. Their size, speed and temperament doesn't make them ideal travelling companions, particularly for a period of several months, but Clara was unusually docile: her upbringing had left her comfortable with human company and she seemed unfazed by this latest episode in her extraordinary life. The crew, no doubt grateful for this, shared their beer rations with her, while her owner, on vet's advice, blew tobacco smoke at her as a prophylactic. It's said she never lost her taste for it, nor for the odd swig of ale.

Clara arrived in Rotterdam in July 1741, the first live rhino to set foot on European soil for more than 150 years. Word spread fast and Captain van der Meer soon realised she could be a major asset on tour. He gave up his job at sea and became her full-time manager, touring Europe with her for the next seventeen years.

Clara sparked two decades of 'rhinomania' as crowds flocked to admire this seemingly prehistoric creature. She was shown off in Hamburg then Brussels, travelling in a carriage, custom built to bear her weight, drawn by twenty horses or six oxen. She weighed over 4,000 pounds and ate more than sixty pounds of hay and twenty pounds of bread a day.

To European audiences, the rhino was almost as mythical as a unicorn – in fact the horn led many to believe that they were actually one and the same. Albrecht Dürer's woodcut image from 1515 added a horn growing out of the upper part of the neck and made the beast look as if it was wearing a suit of metal armour.

From 1746 Clara travelled extensively throughout Prussia, Italy, Switzerland, France and Austria, visiting London a number of times, where she was viewed by the royal family. She was a magnet for artists, many of whom painted her, and she was immortalised in decorations on mantle clocks and medals, and in songs and poems. By the end of the decade cheering crowds flocked to see her wherever she appeared, something she accepted with her usual calm.

Clara in Vienna, engraving of 1746.

She gave private audiences to the great and the good, including King Frederick II of Prussia, Augustus III of Poland, Empress Maria-Theresa of Austria and King Louis XV, who turned down the chance to buy her because the asking price was too high. The women of Paris started wearing their hair in a horn-like shape, styled *à la rhinoceros*, and at Versailles courtiers sported rhino snuffboxes.

As rhinomania spread across Europe, van der Meer capitalised on his protégée's fame by commissioning a range of souvenirs, including engravings and tin medallions. Pictures of Clara adorned porcelain and other luxury goods. Even horses were festooned with rhino-themed feathers and ribbons to keep up with the trend.

Those who saw her, those who wished they had and those who heard about her were universally enchanted and enlightened. Clara's appearance on the Continent changed the popular perception of the rhinoceros once and for all.

Poaching and development meant the Indian rhino faced extinction around a century ago, with fewer than 200 left in the wild. Now, thanks to immense efforts in conservation, there are around 2,600. The total number of rhino worldwide (including Black and White) is thought to be 27,000–30,000.

CONGO

Congo's talent was discovered by Desmond Morris in 1956 while he was conducting a three-year study at London Zoo as part of his research for *The Naked Ape: A Zoologist's Study of the Human Animal*. The first time Morris gave the chimp a pencil and a piece of card, he noted, 'Something strange was coming out of the end of the pencil. It was Congo's first line. It wandered a short way and stopped. Would it happen again? Yes, it did, and again and again.' Morris was amazed by what he saw. 'No other apes were controlling the mark, making and varying the patterns as he was', he explained. 'Congo was a genius. He was the Leonardo of chimp painting.'

Congo concentrating on his painting at London Zoo, 1958.

When Morris gave Congo paints, 'initially it was splish splosh with no direction. He would take the colours I gave him in pots and mix them up into brown. So I started to give him pots of colour in random order.' This seemed to do the trick. 'He was experimenting with forms, especially the form of a fan, balancing compositions, creating repeated motifs and experimenting with colour juxtaposition.'

The chimp, whose style was described as abstract impressionism, continued to progress. He clearly enjoyed painting and was prolific, completing over 400 works. He was very possessive about his art – if someone tried to take a piece away before he had finished, he would (apologies) go 'ape'.

As any creative knows, good PR is essential, and Congo had the benefit of a high TV profile. He appeared regularly on Morris's programme from London Zoo called *Zoo Time* and became the envy of many a contemporary artist by selling work in his own lifetime. An exhibition of his work at the Institute of Contemporary Arts in London in 1957 sold many pieces. On seeing one of Congo's paintings, Salvador Dalí uttered, 'The hand of the chimpanzee is quasi-human. The hand of Jackson Pollock is totally animal.' Picasso and Miró were also admirers of this great artist; the former is said to have hung one of Congo's canvasses in his studio.

After his death in 1964 from tuberculosis, Congo's work became even more valuable. In 2005 three of his works were auctioned by Bonhams for over £20,000 and proved more popular than lots of Renoir and Andy Warhol. And, in 2019, fifty-five of his paintings and drawings (taken from Morris's own collection) were exhibited in a solo show at one of London's leading galleries, priced between £1,500 and £6,000 apiece.

Morris believes that Congo had an aesthetic sense and was

very controlled in his creation of abstract pattern. 'Watching him paint was like watching the birth of art', he said.

CROMMIE

Imagine discovering that your beloved family pet had a hidden heroic past. For Charles Shaw, Cromwell (known as Crommie) was simply 'the first in a long line of wonderful pets throughout my life'. It was not until 2015, when he was in his late sixties, that he learnt the extraordinary truth about his golden Cocker Spaniel.

During the Second World War, his father, Squadron Leader Cautley Nasmyth Shaw, was involved in undercover work with the Secret Intelligence Services (SIS) at Bletchley Park. Setting up and commanding 'a hush-hush unit at Farm Hall, Godmanchester near Huntingdon, to train secret agents', and separated from his family, he took his new dog along to keep him company. Farm Hall was a holding centre – a safe house for despatching agents from airfields at Tempsford or Harrington, or debriefing them as they returned from their assignments.

As the agents prepared to fly into Nazi-occupied Europe at the height of the Second World War, they knew that there was a very strong chance that they would never make it home again. Their fear before these missions was palpable: if caught, the men faced certain torture and death. Their missions were so top secret that they couldn't share how they felt with another living being. At least, not one who could talk. That's where Crommie came in – to offer much-needed calm and comfort.

The SIS which, during the course of the war, became MI6 (standing for section 6 of the Military Intelligence Service) was considered the most effective intelligence service in the world, undertaking missions in Europe, Latin America and Asia. The pressure on their agents – known as Joes – was immense.

Documents belonging to the squadron leader's deputy, Wing Commander Bruce Bonsey, reveal how much difference the dog was able to make to the men. On one occasion, a Czech agent was forced to return to base after bad weather in the drop zone meant he couldn't land as planned. Having geared himself up to go, he was hugely distressed at not being able to complete his mission. Bonsey wrote that the moment he saw Crommie he burst into tears, picked him up and refused to put him down all night:

> After arrival at Farm Hall, he refused a meal or a drink, went straight upstairs, put the dog on his bed, got undressed, took Crommie in his arms and went to sleep. The Conducting Officer told me that the next morning when the Joe woke up and found Crommie still there, he was like a new man, happy, calm and flat out to go off again that night.

His mission is believed to have been successful.

Bonsey described Crommie as 'a Golden Cocker Spaniel of unusual intelligence, and a very happy, friendly nature'. He became so well known amongst the special agents that his name was even used as a vital code word to indicate that they had arrived in France safely: 'love to Crommie' indicated that all was well.

Crommie's calm and sociable temperament was the perfect antidote to the stress heaped on every spy before they undertook a mission. Bonsey saw him as a secret weapon: 'Many times I saw him go and sit beside a Joe who was worried or unhappy, with remarkable results in cheering the chap up with this sympathetic attention.'

Animals had already been proven to help in difficult circumstances. Florence Nightingale noted that a small pet 'is often an excellent companion for the sick, for long chronic cases especially'. (She famously had a pet owl, Athena, as her constant companion, travelling everywhere in her pocket.) Sigmund Freud believed dogs were able to sense and alleviate tension, and that the presence of his dog Jofi in psychotherapy sessions made it easier for his patients to open up. Recent research has shown that in the presence of dogs, a person's level of cortisol drops, resulting in the reduction of anxiety. Even though Crommie had received no training, he had an innate ability to sense when people needed his help. That made him invaluable. As Charles says, 'He was just a normal dog – but for a normal dog he was just amazing.'

When the war ended, Crommie returned to the family farm in Africa and settled into life as a much-loved pet. He died peacefully at a ripe old age in 1953 or 54. 'I was only five or six years old and had no idea of his illustrious past.'

It was not until the discovery of Wing Commander Bonsey's papers that Charles Shaw finally learnt the extraordinary truth about Crommie. When, in 2019, their beloved dog was posthumously nominated for a PDSA Commendation, Charles and his brother David received it on Crommie's behalf.

At the ceremony, Amy Dickin who oversees the awards, described Crommie as 'a forerunner to what we now know as therapy dogs . . . but doing that job without even knowing it'.

David Shaw paid tribute to those who worked behind the scenes at Hall Farm and

> who contributed to the wellbeing of the 'Joes', the moral component, the supply and sustenance of the will to fight, the morale boosters. Now we know that Crommie was part of that component, providing a calmness, a routine presence and a completely non-judgemental devotion that was truly inspiring – all, actually, while being a so-called 'normal dog', quite amazing.

Afterwards, a statue was unveiled at Farm Hall, with a blue plaque which reads 'Crommie. Who comforted agents during World War 2.' Charles added,

> It was a privilege to be part of a commemoration of courage, commitment and intrepid derring-do in the dark days of war when our island nation was fighting for its very survival. They say that 'every dog has its day': well, Crommie certainly had his on that day.

DAISY

Psychologist and animal behaviourist Dr Claire Guest suffers from a rare condition called prosopagnosia which means she is unable to recognise human faces. As a child, she at times found life bewildering, but watching animals proved hugely therapeutic and she could recognise different dogs.

When she went to study psychology in Swansea, she was fascinated by studies into the effects of pets on humans. Her first job was with a charity that trained dogs to help people who were deaf or hard of hearing. It was there that a conversation with a colleague was to change her life.

Guest's co-worker told her how her pet Dalmatian had constantly licked and sniffed at a small mole on her calf, even when she was wearing trousers. The dog's persistence eventually led her to visit her GP. The mole proved to be a malignant melanoma. Claire was inspired by what she had heard. She wanted to prove what a difference dogs could make in terms of saving human lives and teamed up with Dr John Church, whose research had resulted in similar success stories.

In 2004, the *British Medical Journal* published a groundbreaking study on dogs who had been trained to find the odour of human bladder cancer in patients' urine. Our genetic scent can be changed by illness and the scent can be collected in urine, sweat or breath, depending on the disease. To train a dog to identify the disease in question, it has to be given infected samples. Of course, the dog also needs the odour of healthy people as a comparison. When the dogs find a positive result (i.e. signs of cancer) they will sit and stare at the sample to draw attention to it. Whenever they do this correctly, they are rewarded. Despite the findings of the *BMJ* study, many scientists remained sceptical that dogs could match the accuracy of an electronic biosensor. Undaunted by the doubters, Guest started a charity called Medical Detection Dogs. She started to train Daisy, her Fox Red Labrador puppy, to detect prostate cancer. A couple of years later, Daisy suddenly began to act strangely around her. She seemed wary and hesitant and one day refused to get out of the boot of the car but kept jumping up at Claire's chest. When eventually Daisy joined her other dogs for a run, Claire absent-mindedly rubbed at the spot that the dog had been indicating. She could feel a lump.

Assuming it was probably nothing, she went to get it checked out just to be on the safe side. Within a fortnight she had been diagnosed with breast cancer.

The lump was benign but the cancer was hidden deeper beneath, close to her heart. By the time Claire would have been able to detect it, the prognosis would have been very different. Daisy's warning had saved her life – something that impressed her doctors so much that Claire's oncologist became a trustee of Medical Detection Dogs.

Daisy went on to identify correctly thousands of cases of cancer and was awarded a Blue Cross Medal for her services. Heartbreakingly she died of cancer herself. Dr Guest was devastated by the loss of the dog who 'had done so much for humans and for me personally'.

Daisy's legacy lives on. Medical Detection Dogs now has thirty-five bio-detection dogs – all of whom go home to live with volunteers at night rather than being kept in kennels. Daisy's niece has done important work on prostate cancer at MIT in America and a great-nephew detects E. coli.

Guest, who is now working with the London School of Hygiene and Tropical Medicine and Durham University, hopes that dogs might be able to have a significant impact in the fight against COVID-19. It could just be that dogs will be the ones able to identify the disease with speed and reliability, even in those who are asymptomatic. That would be a game changer.

DIESEL

I am lost in admiration for the dogs who work in emergency situations. Whether they are catching criminals, finding drugs or patrolling sensitive areas, they are high-value members of the forces. At Crufts every year there are demonstrations in the main arena of dogs used by the police, the RAF and the army. It is an education for the crowd and has helped me gain a greater level of understanding of what they do and how much more they can achieve than humans or technology. Dogs love to learn and they all have their own reward scheme – often it's the chance to play with a toy rather than a treat. The dogs that I have seen at work genuinely enjoy what they are doing and have no awareness of the potential dangers.

The work of a RAID (Research, Assistance, Intervention, Deterrence) assault dog is complex and highly specialised: it takes months of training for them to learn to identify the many explosive compounds and devices that they may need to seek out in the course of their anti-terrorist duties. They are naturally well equipped: dogs' sense of smell is up to fifty times better than that of humans. They are agile and fast, brave and focussed but their duties are a matter of life or death for them and all the people around them.

So it proved for Diesel, a Belgian Malinois, who had spent five of her seven years working for the French police before she hit the headlines on 18 November 2015.

Just five days earlier, the French capital had been rocked

by an appalling series of terror attacks. It began at 9.20pm on 13 November with a suicide bomb outside the Stade de France, where France were playing Germany in a friendly. It ended just after midnight at the Bataclan concert hall. In between, suicide bombers and gunmen caused carnage in restaurants and bars filled with people enjoying a Friday night out. In just a few hours, 130 would be dead and 350 injured, many critically.

On 18 November, police raided a flat in the St Denis area of Paris, searching for Abdelhamid Abaaoud and others suspected to be behind the massacre. It was a violent encounter, with up to 5,000 bullets fired and dozens of grenades thrown. Suddenly it all went quiet and Diesel was sent in to sniff out the suspects. The first room was empty, but as she entered the second, the shooting began again. The dog was caught in the crossfire. She suffered multiple gunshot wounds and did not survive. She became the force's first dog to be killed in active service.

Parisian police chiefs told the media that Diesel had almost certainly saved her handler's life during the raid, and tributes to the dog's bravery went viral on social media, under the hashtags #JeSuisChien and #JeSuisDiesel, echoes of the tags used after the Charlie Hebdo attacks in January that year.

Diesel had been just months away from retirement and her story touched the world. The Russian interior ministry sent a replacement puppy to show their solidarity with the people of France. And on 28 December, Diesel was posthumously awarded the Dickin Medal for her bravery.

DOLLY

It's not often you can find a sheep who is prepared to be different and I guess even Dolly was unique because she was the same. She was the first mammal ever cloned. Cloning is now standard practice in some countries. In South Korea you can get pets cloned if you've got enough money – around £100,000 – and in Texas there's a company which will clone your cat for \$25,000 and your dog for \$50,000. In Argentina, there is a line of six clones of the same outstanding polo pony, all with the same name and a different number.

This all started with a sheep, an animal widely written off as being stupid. Sheep are in fact intelligent and have excellent memories. True, they follow the crowd, but that is because they know that standing out leaves them vulnerable to predators. They are sociable and can form firm friendships. They have great powers of recognition: a study in 2001 found that they can recognise and remember at least fifty individual faces for more than two years – longer than many humans. Nonetheless, it would be fair to say that one wouldn't necessarily expect a sheep to be at the forefront of scientific discovery.

That's why I have singled out Dolly as a game changer for her contribution to the world of science.

Born in July 1996, Dolly the Finn Dorset lamb was the first mammal ever successfully cloned from an adult cell. Cloning is the process of producing genetically identical individuals of an organism either naturally or artificially. In Dolly's case, scientists achieved this using 'somatic cell nuclear transfer'. This is a process whereby the nucleus from a mammary gland cell of one organism is transplanted into the egg of another, which has had its nucleus removed. Electricity is then used to fuse the two together and stimulate cell division. In this way, a whole individual can be created from one cell taken from the specific part of a body.

Asexual reproduction had been achieved in plants through grafting and stem cuttings for more than 2,000 years, but it only moved into the laboratory and became significantly more high tech in 1958, when the botanist F. C. Steward took mature single cells and placed them in a nutrient culture containing hormones. The result: the first cloned carrot plants.

Six years later came the first experiment using animal cells. Biologist John Gurdon injected the nuclei from intestinal cells of toad tadpoles into unfertilised eggs whose nuclei had been destroyed using ultraviolet light. Between 1 and 2 per cent of the eggs developed into fertile, adult toads. Many other trials were to follow, and Dolly proved to be the most significant breakthrough of all.

Her birth was only announced some months afterwards, in February 1997. Unsurprisingly, it received a lot of attention. Her story was featured in *Time* and *Science* magazines, and BBC News called her 'the world's most famous sheep'.

Despite her high-profile status, Dolly lived a quiet and happy life at the Roslin Institute, where she was mated with a Welsh Mountain ram and produced Bonnie (1998), twins Sally and Rose (1999) and triplets Lucy, Darcy and Cotton (2000).

Finn Dorset sheep generally live to eleven or twelve, but Dolly suffered from severe arthritis and a progressive lung disease and died aged six and a half in February 2003. Some critics suggested that she died halfway through her breed's natural lifespan because she was born with a genetic age of six (the age of the sheep from which she had been cloned). Further investigation saw scientists report no evidence of detrimental long-term effects, and thirteen more cloned sheep, among them four from the same cell line as Dolly, showed no defects.

Cloning may continue to have its critics, but the technology could also have widespread benefits. It may help to preserve endangered species, even to reverse the extinction process where frozen tissue is available. In 2009, scientists in northern Spain managed to resurrect the Pyrenean ibex by cloning. The species had been declared extinct almost a decade earlier.

More recently, researchers in China have achieved the first successful cloning of a primate species using the Dolly method. Two identical clones of a macaque were born in 2017, with more monkeys created in 2019 to help scientists study a number of medical conditions. Cloning has also contributed significantly to stem cell research, something that *Scientific American* pinpointed as being perhaps the greatest legacy left by Dolly.

DYLAN

Tiny birds can become big heroes, as Andy Hardiek of Avilla, Indiana, discovered in January 2014.

Exhausted after a long night shift at the factory where he

worked, thirty-six-year-old Hardiek was fast asleep when fire broke out beneath his mobile home. As noxious smoke and flames began to fill the trailer, Dylan, his pet cockatiel, did everything he could to alert his owner, jumping up and down, rattling his cage and squawking loudly.

Hardiek finally stirred from his slumbers. He grabbed a fire extinguisher to put out the blaze but realised that the flames had already taken hold, so he picked up Dylan's cage and ran to safety. Their home (and all their possessions) were destroyed, but the pair escaped unharmed.

Hardiek told a local news station, 'Without Dylan waking me up, I probably wouldn't be here right now', and the local fire chief agreed: 'The bird notified him before the smoke detectors. I've heard of dogs and other animals waking people up from fires, but never a bird.'

FÉLICETTE

At the peak of the Cold War, the world's superpowers were obsessed with going into space. The race between the USA and the Soviet Union saw a dog called Laika (of whom more later) orbit the earth in 1957 among other animal astronauts. The desire to break new frontiers was infectious, leading other countries to join the battle for a bite of the intergalactic cherry.

At the forefront was France, which has the world's third oldest institutional space programme. The Centre National d'Études Spatiales (CNES) was formed in December 1961, and these days it has the largest national budget for civilian space programmes after NASA. Back then, at the start of their story, they wanted to send a cat to boldly go where no cat had gone before (or indeed since).

The safety record of animals in space was mixed. Fruit flies launched in a rocket in February 1947 returned safely (using a parachute) after reaching an altitude of sixty-eight miles, but experiments using monkeys were catastrophic. Albert, the first primate launched, died of suffocation before his rocket reached peak altitude. In June 1949, Albert II became the first simian in space, reaching an altitude of eighty-three miles, but on his return to earth his parachute failed. He died on impact. The fates of Albert III and IV were much the same. Parachute failure was also behind the death of the first mouse in space in 1950, when her rocket disintegrated. Dogs, as we'll discuss later, had a similarly grim record so things did not bode well for the cat.

Félicette was one of fourteen cats purchased from a Parisian pet dealer to be trained for the job in hand. They were selected according to their temperament, and all the cats were female because they have a calmer outlook on life. To reduce the chance of any scientists becoming emotionally attached to them, all the cats were unnamed prior to the launch. One of the fourteen – known unromantically as C341 – was selected. Some say it was because she was the most docile, other reports suggest that it was because all her fellow felines had put on too much weight.

On 18 October 1963, she was launched in a rocket from the Sahara Desert to fly almost 100 miles above the earth. She underwent the full astrocat experience: G force as she soared at up to six times the speed of sound, a spot of weightlessness and then, after eight minutes and fifty-five seconds, a capsule detached from the rocket and parachuted her back to earth. She was retrieved by helicopter thirteen minutes after launch, safe and well.

Félicette's signed photo.

After the flight, the media christened her Félix (Latin for lucky) after the celebrated cartoon cat, but staff felt the female form was probably more appropriate, and so Félicette she became. Before Félicette, France had only launched rats into space. Now it had something to catch them and certainly to generate more press interest in the space race.

Félicette's life was high profile but short. A few months later, she was euthanised so that scientists could examine her brain and her body to see what effects her flight in space may have had.

Her contribution to increasing the status of France in the space race took a long time to be recognised. In 1992, the former French colony in the Comoro Islands featured her on a stamp, as part of a set celebrating animals in space. Then in 2017, some fifty-four years after her epic space flight, a crowd-funding campaign was started to erect a memorial to her. Two years later, in 2019, a bronze statue of Félicette was unveiled at the International Space University in Illkirch-Graffenstaden, France.

The one and only astrocat, Félicette now perches on a model of Earth, gazing up towards the skies she once travelled. It may have been one small step for a cat but it was one giant leap for catkind.

FINN

There is, perhaps, no greater act of heroism than putting the life of someone you love ahead of your own, but this is exactly what Finn, a German Shepherd, did in the autumn of 2016.

PC Dave Wardell, an officer with the Bedfordshire, Cambridgeshire and Hertfordshire Police Dog Unit, was called out with Finn on 5 October to search for an armed man following a robbery in Stevenage.

As Wardell and Finn tracked him down, the suspect made a break for it and tried to escape by scaling a nearby fence. Wardell shouted a warning to him to stop, but it fell on deaf ears. This was Finn's moment – could he stop the youth before he managed to get away?

As Wardell released him, the dog made straight for the suspect, seizing his leg and bringing him to the ground. The youth drew a twelve-inch knife and stabbed Finn in the chest, narrowly missing his heart. Wardell rushed to intervene, but the suspect continued his attack, lashing out brutally, this time slashing the dog's head and his owner's hand.

Finn was gravely wounded but refused to release his grip, allowing Wardell to disarm the suspect without sustaining further injury. Help arrived and the dog was rushed to a vet to undergo emergency surgery. The damage sustained in the attack was so severe that part of his lung had to be removed. He almost died. 'Finn was linked up to all sorts of machines and breathing aids', said an emotional Wardell. 'All I could see was my big, brave boy so horribly diminished. Almost his whole body was shaved, beneath a blue protective jacket that kept all the tubes in place and protected his enormous surgical wounds.'

Finn made a remarkable recovery, returning to active duty in just eleven weeks. He retired shortly before his eighth birthday in March 2017 and was happily adopted by his handler.

Two months later, his attacker was convicted of assault and causing actual bodily harm at Stevenage Youth Court, but his crime only covered the wounds he had inflicted on PC Wardell. Those suffered by Finn could only be classed as 'criminal damage', as if he had damaged a car or smashed a window. He was sentenced to eight months in a young offenders' institution.

Animal lovers were outraged, and thousands of them supported PC Wardell's attempts to change the law to increase sentences for those who abuse animals, and for UK law properly to recognise animals working for the emergency services. Over 127,000 people signed an online petition campaigning for 'Finn's Law'. This led to recommendations that any future attacks on emergency service animals be treated as an 'aggravated offence'.

Sir Oliver Heald, MP for North East Hertfordshire, PC Wardell and Finn's constituency, took things a step further by proposing a private member's bill, which was debated in the Commons in December 2017. It received unanimous all-party backing and then Royal Assent in April 2019.

The Animal Welfare (Service Animals) Act 2019 came into force two months later, offering service animals increased protection by amending section 4 of the Animal Welfare Act of 2006. In Scotland, the First Minister, Nicola Sturgeon, announced that Finn's Law would be incorporated into a new Animal Welfare Bill as part of a new programme of legislation to be introduced into the Scottish Parliament. At the Northern Ireland Assembly, a motion to incorporate Finn's Law was passed unanimously in February 2020 and passed to their Agriculture Minister Edwin Poots.

What of our hero himself?

It has not been a quiet retirement. The media coverage around the time of the attack brought his bravery to the attention of the general public and he was also recognised more formally with a number of honours. These included the International Fund for Animal Welfare Animal of the Year Award in October 2017 and the PDSA Gold Medal 'for life-saving devotion to duty, despite being grievously injured while preventing a violent criminal from evading arrest' in May 2018. Finn was also given the Kennel Club's 'Friends for Life' award at Crufts in March 2019.

The strength of the bond between Finn and Dave Wardell was illustrated perfectly in the spring of 2019, when the pair wowed the audience of *Britain's Got Talent* with an extraordinary mind-reading act that left everyone watching in tears and took the pair right up to the final.

Wardell asked the judges to think of a word and write it

down. David Walliams picked 'table', keeping the card close to his chest so that Wardell couldn't see what he had chosen. After the judge had shown the card to Finn, Wardell called his dog up on stage and knelt beside him. The audience watched spellbound as Finn appeared to whisper something into his master's ear. Wardell told the judges that the word was 'table', then stood aside as the screen behind him showed a video he had put together of Finn's story. There was barely a dry eye in the house. Simon Cowell wept openly, telling Wardell, 'If I had a golden buzzer left, I'd give it to you. When I hear about animal cruelty, especially dogs, it upsets me. A dog will literally give up its life for you. Finn's beautiful, I love him.'

Finn will be remembered for far more than his moment of TV stardom. The change in the law brought about by his exceptional bravery makes it a criminal offence to harm or abuse any animal operating in the line of duty. That's quite a legacy.

FRANKIE

Dogs are known for their exceptional sense of smell, which is fifty times better than any human's. They can sniff out truffles from deep underground, track down criminals, seek out explosives and even detect bed bugs. Their ability is employed by customs officers to uncover drugs or other prohibited substances. People suffering from diabetes can be alerted to falling blood sugar by dogs that are able to detect a specific scent on their breath, and dogs can predict an epileptic fit.

Not all of them can do this, of course. My own dog, Archie,

a Tibetan Terrier, did not show any signs of awareness when I developed a large lump on my throat. The GP checked it out and sent me to an ENT specialist who recommended a biopsy. The results revealed I had thyroid cancer. I had three operations to remove the thyroid and a bout of radioactive iodine treatment to destroy any remaining fragments. Touch wood, I've been clear ever since but if I'd met Frankie in 2009, he might have warned me sooner.

The German Shepherd mix stray from Little Rock became the star of a study by researchers at the Arkansas University for Medical Sciences (UAMS). By smelling patients' urine samples, he was able to tell with 88 per cent accuracy which tumours were malignant and which benign.

Frankie was at the forefront of new research into the use of dogs to diagnose thyroid cancer through scent imprinting, the process that is also used in training dogs to sniff out explosives and drugs. By learning the traits of a specific smell, the dogs are able to identify it whenever they come across it in future. Previous studies had already shown that dogs had the ability to pick out urine samples of patients with cancer from those given by people in full health. Now it was time to get specific.

For six months Frankie was trained using samples from patients with cancerous thyroid growths. Constant exposure to their blood, urine and other tissue meant that he was then able to differentiate between metastatic carcinomas and benign tumours. If he spotted the latter he would turn away, but if the samples showed malignancy he would lie down to indicate a positive result. Thirty-four patients were tested in the trial and Frankie got the correct diagnosis thirty times.

Using dogs to help advance diagnosis of potentially fatal

diseases makes absolute sense when you consider the physiological advantages they have over humans. We breathe and smell through the same airways, whereas dogs' noses have a flap of skin inside the nostril, which partly separates the functions. The part of their brain which analyses smell is proportionately much bigger than ours. Dogs also have an active VNO, or vomeronasal organ, in their nasal septum, which picks up pheromones and other chemicals that stimulate physiological and/or behavioural changes. This means that dogs can literally smell fear, anxiety and other emotional states in a way a human never could.

Alexandra Horowitz, author of *Inside of a Dog: What Dogs See, Smell, and Know*, says: 'As it turns out, humanity's best friend is not one who experiences the same things we do but one whose incredible nose reveals a whole other world beyond our eyes.'

GALLIPOLI MURPHY

The stories told about the First World War concentrate almost entirely on those set in rat-infested French and Belgian trenches, but in fact terrible battles were fought across the globe. One of the worst was the Gallipoli campaign where a donkey called Murphy was the saviour of many wounded men.

On 25 April 1915, soldiers from Australia and New Zealand landed at Gallipoli Cove in Turkey, joining the British and French effort to force pro-German Turkey out of the war. The plan was to capture the peninsula then move inland, but the Turks fought back hard and the Allied troops were trapped near the beaches where they had landed. Around 650 men were killed that day alone, with more than 1,000 others wounded. It was a bloodbath.

Among the soldiers landing at Gallipoli was John Simpson, a stretcher bearer with the 3rd Australian Field Ambulance. Simpson's job was to provide first aid to the wounded and then to carry them to the beach, from where they could be evacuated.

John Simpson had had a chequered past. Born John Kirkpatrick in South Shields to Scottish parents, he had always loved animals and worked his school holidays giving donkey rides on the beach. He trained as a gunner in the Territorial Force at sixteen and joined the British Merchant Navy. But it was not the life for him, and he hated it so much that in May 1910, when the ship docked in New South Wales, Australia, he got off and never went back.

He travelled through Australia, picking up a range of jobs including gold panning, cane cutting and coal mining. The outbreak of the First World War gave him the opportunity to reinvent himself so, using his mother's maiden name of Simpson (in case anyone realised he was a deserter), he joined the Australian and New Zealand Army Corps (ANZAC).

So it was that he found himself on the beach at Gallipoli in the early hours of 26 April, carrying a wounded man on his shoulders towards the first-aid station. That was when he spotted a donkey.

Donkeys had long been the unsung heroes of the battlefield. During the First World War, long trains of them, 200 at a time, were used to bring vital supplies to the troops. They were routinely loaded with at least three times their own body weight. Usually under cover of darkness, they carried food supplies, clothing, pots and pans, and – crucially – water while all around them guns fired and bombs exploded.

Simpson realised that it would be far easier to transport casualties with the donkey's help. They became a vital life-saving team. Everyone knew about him and would ask, 'Has the bloke with the donk stopped yet?' He didn't stop and nor did Murphy the donkey. Captain C. Longmore, interviewed in 1933, remembered how the soldiers had 'watched him spellbound from the trenches . . . it was one of the most inspiring sights of those early Gallipoli days.'

John Simpson with Murphy who carries a wounded soldier, 1915.

During his time at Gallipoli, some say that Simpson worked with more than one donkey: the first known as Duffy, and then Jenny, Queen Elizabeth and Abdul before the last one, Murphy. That seems likely as their work was extremely perilous and they were under constant fire with bullets flying every which way. Other accounts say the donkeys were one and the same, plucky little Murphy, who gained legendary status and became a hugely popular mascot to the ANZACs.

News of their bravery soon reached Simpson's seniors. One, Colonel – later General – John Monash wrote: 'Private Simpson and his little beast earned the admiration of everyone at the upper end of the valley. They worked all day and night and the help rendered to the wounded was invaluable.' The military were so impressed by the way the donkey had managed quickly to convey the wounded across the broken and dangerous ground, that large groups of donkeys, all sporting Red Cross headbands, were after that held in readiness to help the stretcher bearers.

John Simpson was killed by machine-gun fire in Shrapnel Valley during the third attack on Anzac Cove, three weeks later on 19 May 1915. He was just twenty-two. He is buried on the beach at Hell Spit at the southern end of the cove.

What happened to Murphy is less clear, but accounts offer a ray of hope. As with Minnie the mule, soldiers often went to enormous lengths to get their pets to safety. Pets on the battlefield gave men a link with home and a reason to keep fighting. They offered something to care for and a welcome change from guns, bombs, lice and dirt. Even officers were known to have them despite rules to the contrary. A letter dated 21 March 1916, addressed to Captain Charles Bean, the official Australian war correspondent who was trying to trace Murphy's fate, states:

You will be pleased to hear that Murphy was safely evacuated . . . but the night he arrived at Mudros [a nearby Greek island], he disappeared and though all the nearby villages were searched, no trace of him was found . . . My belief is that he was taken from the lines by Australians as he had on him two huge labels on which were printed 'Murphy VC: please look after him.' I hope it is so and that he is now in their possession. My men begged of me to get him away.

In 1997, the RSPCA Purple Cross and certificate of award was retrospectively given to Murphy the donkey, in recognition of his service.

GANDER

Newfoundland dogs are known for their loyalty and calm disposition, but they are big beasts. They stand at over two feet tall and weigh up to ten and a half stone. They also drool a lot. Newfie owners adore their pets but they are not suitable for everyone.

Pal, owned by the Hayden family, was a favourite Newfie with the local youngsters, often pulling a sledge for them in the winter. He was a gentle giant but totally unaware of his own strength. On one occasion he accidentally scratched the face of a six-year-old girl and when the doctor was called to treat her wound, the Haydens were terrified they would be asked to have him put down.

They decided to hand the dog over to the Royal Rifles of Canada, a Canadian Army regiment stationed at nearby

Gander International Airport, Newfoundland and Labrador. There, Pal, renamed Gander, rapidly became a favourite with the soldiers, who soon 'promoted' him to sergeant and took him along with the rest of the unit when they were shipped out in the autumn of 1941 to defend Hong Kong from enemy invasion.

Men of the Royal Rifles of Canada with their mascot Gander, on their way to Hong Kong in 1941.

Hong Kong's tropical climate was somewhat different from the Canadian cold that Gander and his comrades were used to. Rifleman Fred Kelly, who was assigned to take care of the dog, reportedly allowed him to take long cold showers to help stave off the worst of the heat. It's said he also became a fan of a cooling beer (or two).

The day after the attack on Pearl Harbor in early December 1941, the bitter Battle of Hong Kong began, one of the first battles of the Pacific War. Japanese forces attacked the British colony, even though their action violated international law, as Japan hadn't officially declared war on the British Empire.

The battle lasted over two weeks, with Japanese troops outnumbering Allied forces by two to one. The losses were accordingly heavy, with over 2,000 British, Canadian and other Allied soldiers losing their lives, 2,300 wounded and 10,000 captured. Four thousand civilians were also killed.

Despite the mortal danger, Gander helped to fight off the invaders on three occasions. He would growl ferociously, running at enemy soldiers and snapping at their heels. His dark fur made him hard to see at night, when much of the fighting took place. Indeed, Japanese soldiers later interrogated Canadian prisoners of war to find out more about the 'black beast'. They were terrified he was part of a force of barbarous quadrupeds being trained by the Allies to attack their enemy.

Sergeant Gander's final act was to save the lives of seven wounded soldiers by removing a grenade. 'Gander must have seen the Japanese hand grenades landing', says Jeremy Swanson of the Canadian War Museum. 'He must have seen the men furiously throwing them back. He must have sensed their terror.' He picked it up and ran with it towards the enemy. Tragically, it exploded before he reached them. The lives of the seven men had been saved, but as Reginald Law remembers, 'When the firing eased up, I saw Gander lying dead in the road.' The next morning, as they were marched away into captivity, Fred Kelly could see Gander's body in the distance: 'I didn't go near, I was so distraught.'

After the war the Canadian War Museum, the Hong Kong Veterans Association and the Hong Kong Veterans

Commemorative Association made every effort to ensure that the Royal Rifles' heroic mascot be recognised for his outstanding deeds. In October 2000, Sergeant Gander was posthumously awarded the PDSA Dickin Medal, the only Canadian recipient ever. The twenty surviving members of his regiment, including Fred Kelly, attended the ceremony. 'It's very emotional, even talking about it, it's very close to my heart', said Philip Doddridge, who had fought alongside Gander, and spent three years in a Japanese prisoner-of-war camp. 'He was very much loved by all of us, he followed us to Hong Kong and was killed in action.' I wonder how many of them were aware that he had once had a very different life.

GEESE

Our story is set in the Roman Republic well before the days of the Empire, on the Capitoline Hill, the smallest yet most significant of the seven hills of Rome. Within the citadel were temples dedicated to the gods, including one in honour of Juno, queen of the gods and goddess of marriage and childbirth.

Juno's temple was guarded by a gaggle of sacred geese. The Romans knew that these birds were to be respected. Geese are very intelligent and remember people and situations. They know the boundaries of their home, which they will protect fiercely. They also know who is friend and who is foe, and can be aggressive if their young are threatened.

In 390 BC, a group of invading Gauls marched towards the

city. They were a brutal armed force, experienced and hardy. The ensuing battle, fought near the Tiber and Allia rivers, proved disastrous for the Romans. Many fled for their lives and large numbers drowned as they tried to cross the river wearing heavy battle armour. The Gauls went on to reach the outskirts of Rome within a day.

Their speed meant that only small numbers of Roman survivors from the battle had made it back to defend their city. Many citizens fled while they could, while the priests and priestesses of the holy temples rushed to remove precious artefacts for safekeeping.

The Gauls stormed the walls, rampaging through the city and killing anyone in their path. Their earlier victories had filled them with confidence, so when they realised that most of the remaining Romans were to be found on the Capitoline Hill, they promptly attacked, expecting to dominate the battle as before.

But the Romans had the advantage of the high ground and fought back. The Gauls retreated to replan their attack but kept the hill under siege. The remaining Romans were stuck and in danger of starving if they could not get help or escape. But somehow provision was made for the sacred geese.

On a night with a full moon the Gauls decided to scale the steep side of the hill. The guards had fallen asleep and the dogs were snoring, so the Gauls were unchecked. Until they came across the geese. The birds realised the intruders were not friends and started to honk and flap and attack for all they were worth.

The noise roused some of the sleeping Romans, among them a man named Marcus Manlius who was first to reach the cliff, where he charged the enemy soldiers reaching the top, killing one and using his shield to knock another tumbling to his death. Still the geese kept up their cacophony, ensuring that other Roman troops were awakened and joined the fight. Any Gaul reaching the top of the cliff was killed. Others, still ascending and clinging to the rocks to survive, stood no chance as they were pelted with stones and javelins until they too fell to their deaths. In the words of Roman historian Plutarch:

> Neither man nor dog was aware of [the Gauls'] approach. But there were some sacred geese near the temple of Juno . . . The creature is naturally sharp of hearing and afraid of every noise, and these, being specially wakeful and restless by reason of their hunger, perceived the approach of the Gauls, dashed at them with loud cries and so waked all the garrison . . . So the Romans escaped out of their peril.

Although the siege continued, it became clear that the Gauls would never take the Capitol. By now they were finding things tough, and both sides reached an agreement that the invaders should be paid to leave the Romans in peace.

Rome was rebuilt and eventually grew to dominate the world. Its citizens never forgot their debt to the geese of Juno, who were remembered every year with a magnificent procession, when a single goose dressed in gold would be carried to the fore, to honour the sacred birds that saved their city.

The Roman Goose, with its tufted topknot, is one of the breeds still used today as a 'guard goose'. Chinese and African geese are other good options to protect chickens and flocks on farms, to patrol military centres and even to guard whisky warehouses in Scotland.

G. I. JOE

Of the seventy-one recipients of the Dickin Medal, the animal equivalent of the Victoria Cross, one has been American. His name was G. I. Joe.

Joe was hatched in Algiers in March 1943 under the name Pigeon USA43SC6390. He had undergone training for two-way homing pigeons in New Jersey before taking on active service later that year. Before long, he found himself in Italy where American planes were preparing to bombard German-occupied Calvi Vecchia, about thirty miles north of Naples, so that the British could march in and capture the defenders. As the troops of the 169th (London) Infantry Brigade closed in, the Germans retreated unexpectedly, allowing the Allied forces to take the town ahead of schedule. In normal circumstances, this would have been good news but not when your allies are about to launch an air raid on the area you've occupied.

Attempts to transmit a message via radio failed. The lives of the British troops and the many inhabitants of Calvi Vecchia lay in the balance. G. I. Joe flew the twenty miles back to the air base in just twenty minutes, arriving in the nick of time, as the planes were about to take off. Around 1,000 people survived because of his speed that day.

The pigeon's heroism earned him the Dickin Medal for 'the most outstanding flight made by a United States homing pigeon in the Second World War'. It was presented at a ceremony at the Tower of London in 1946.

G. I. Joe and his Dickin Medal at the Tower of London, 1946.

After the war, Joe returned to Fort Monmouth, New Jersey, where he shared the US Army's Churchill Loft with a number of other pigeons who had shown great bravery during battle. He died at the grand old age of eighteen in the Detroit Zoological Gardens. His body was returned to Fort Monmouth, where it was mounted and displayed at the US Army Communications Electronics Museum.

GREYFRIARS BOBBY

Sometimes our best-loved stories grow and change over the years into the ones we need to hear. Perhaps the most famous Scottish tale of heroic canine loyalty has been embellished over the years with a little fictional poetry but I had to include it here.

Dogs are often described as man's best friend and no dog has lived up to this title more than Greyfriars Bobby, a Skye Terrier born in Edinburgh in 1855. He adored his owner John Gray, and never left his side. Gray, or Auld Jock as he was known, worked as a night watchman for Edinburgh City Police and would keep watch with Bobby by his side.

In February 1858, Auld Jock contracted tuberculosis and died.

Bobby led the funeral procession and when Auld Jock was buried at Greyfriars Kirkyard in the centre of the city, he refused to leave his graveside. The caretaker tried to shoo him away, but Bobby returned to sit by his master's side, guarding the grave. It became a habit. Day and night, Bobby watched over Auld Jock's final resting place. Even the worst of Scotland's snow or winter rains could not keep the dog away.

His loyalty touched the local people and, although dogs were not actually permitted in the kirkyard, they built a special shelter so that he could continue to guard his master's body with some protection from the elements.

Bobby left the graveside only once a day, to eat. When the one-o'clock gun fired from Edinburgh Castle, that was the signal for him to go and get his daily meal from the place where he and Auld Jock used to eat. Other than that, he refused to move from his post.

Word of the little dog's loyalty soon spread, and crowds would gather to watch him head off for his lunch and then return to the graveside. The Lord Provost of Edinburgh, Sir William Chambers, was touched by the terrier's devotion. In 1867, nine years after Auld Jock's death, he paid for the dog's licence himself and presented Bobby with a new collar, inscribed with the words, 'Greyfriars Bobby – from the Lord Provost, 1867, licensed'.

Bobby's vigil continued for fourteen years, until his own death in 1872. He was buried at Greyfriars, just yards from his master, and a granite headstone, unveiled by the Duke of Gloucester in 1981, reads:

GREYFRIARS BOBBY

– DIED 14TH JANUARY 1872 –

AGED 16 YEARS

Let his loyalty and devotion be a lesson to us all

His story proved to be an inspiration for the philanthropist Lady Angela Burdett-Coutts. Not long after Bobby's death, she commissioned a statue of him from sculptor William Brodie. It was erected in 1873 above a drinking fountain – an upper fountain for humans and a lower fountain for dogs – standing opposite the entrance to the churchyard at the junction of George IV Bridge and Candlemaker Row. While health scares saw the water supply cut off in 1957, the monument was restored in 1985 and is now Edinburgh's smallest listed building. A plaque reads, 'A tribute to the affectionate fidelity of Greyfriars Bobby'.

Tourists come from far and wide to see the statue of the Skye Terrier and often go on a tour with the Greyfriars Bobby Walking Theatre. There have been films made about Bobby and toys made of him, and the statue has become a symbol of good luck as well as loyalty. In fact, his nose has been so worn down by travellers rubbing it that it's had to be restored twice.

Statue of Greyfriars Bobby by William Brodie.

Bobby's is not the only documented case of a dog showing unswerving loyalty to a deceased owner.

In November 1941, Carlo Soriani, a brick worker in Borgo San Lorenzo, Tuscany, came across a dog lying injured in a ditch. The war meant that many pets had been separated from their families and forced to survive on the streets. Soriani took the stray home and nursed him back to health.

The dog was so grateful to his saviour that he followed him to the bus stop every morning, earning the nickname Fido, Latin for faithful one. Every evening, he would return to sit and wait for Soriani's bus to arrive. This continued for two years until December 1943, when Allied bombing raids reduced many of the area's factories to rubble, killing thousands of civilians, including Soriani.

That night, Fido remained at the bus stop waiting for his master. Every evening for fourteen days, he waited for the bus from which Soriani never emerged.

Before long, people began to notice what was happening and the press picked up on the story of Fido waiting for his owner. After the dog died, in June 1958, a statue was commissioned in his honour.

GRIP AND GRIP

I can't be the only one who finds ravens a bit spooky, but there are real fans out there who will argue that they are the best birds of the lot. They're certainly cunning, versatile and even seem to have a sense of humour.

Ravens are the largest members of the crow family and with a wingspan of over four feet, they are an impressive sight in flight. Black feathers, piercing black eyes and a large, hooked black beak make them pretty intimidating; across Europe they have long been associated with death.

They can also form attachments to humans. Grip, a cheeky and talkative raven, was the much-loved pet of Charles Dickens who made him a character in his novel *Barnaby Rudge*. He wrote to his friend George Cattermole in January 1841:

My notion is to have [Barnaby Rudge] always in company with a pet raven, who is immeasurably more knowing than himself.

To this end I have been studying my bird, and think I could make a very queer character of him.

That same month Grip met a sad end. It seems he may have eaten some white paint from a tin. In a letter to the painter Daniel Maclise, Dickens wrote that even though a vet had dosed the bird with castor oil, which revived him enough for him to bite the coachman and manage some warm gruel, his recovery was short-lived:

On the clock striking twelve he appeared slightly agitated . . . [he] staggered, exclaimed 'Halloa old girl' (his favourite expression) and died. He behaved throughout with a decent fortitude, equanimity and self-possession, which cannot be too much admired . . . The children seem rather glad of it. He bit their ankles. But that was play.

Dickens's raven hero in *Barnaby Rudge* went down very well with the Gothic writer and poet Edgar Allan Poe, who wrote in a review that he found the bird 'intensely amusing'. Grip may well have inspired Poe's famously dark poem 'The Raven', in which a raven conveys a message of endless sorrow upon the narrator with the repeated single word, 'Nevermore'.

Another famous raven, named Grip after Dickens's pet, was one of the ravens guarding the Tower of London, the only Tower raven to survive the Blitz.

There have been ravens at the Tower for hundreds of years. According to one of several legends, John Flamsteed, an astronomer, complained to King Charles II that the birds were interfering with the work of the Royal Observatory in the northeastern turret of the White (or central) Tower.

When the King ordered that the birds should be destroyed, he was warned that if they left, the White Tower would fall and bring disaster to the kingdom. It didn't seem worth the risk, so the ravens received a stay of execution, the observatory moved to Greenwich and the monarch decreed that at least six birds must be kept in the Tower at all times, leading to a belief that without the ravens 'the Crown will fall and Britain with it'.

The myth may be no more than that, but during the Second World War the ravens displayed an uncanny knack for predicting bombing raids. However, the noise of bombing was too much for them and they are said to have died from shock. Only one remained – our hero Grip, who guarded the Tower alone, even after his beloved mate Mabel disappeared (rumoured to have been kidnapped) – until Winston Churchill ordered that the flock should be increased to a minimum of six.

To this day, at least six of the birds are kept at the Tower, enlisted as official soldiers of the kingdom. Mischievous birds, they can be discharged for unacceptable conduct in the same way as any other member of the forces. One, named George, lost his royal appointment and was banished to a zoo in Wales after attacking and destroying TV aerials.

The current group of ravens guarding the Tower are a better-behaved mob. They include Jubilee, Rocky, Poppy and Gripp with two 'p's. They are the product of a special breeding programme and are looked after by an official ravenmaster who feeds them fresh meat including liver, lamb and pork, biscuits soaked in blood with a boiled egg (once a week) and the occasional rabbit, fur and all.

Visitors are advised to keep well away from the ravens, as they will not hesitate to bite. But despite their diva-like

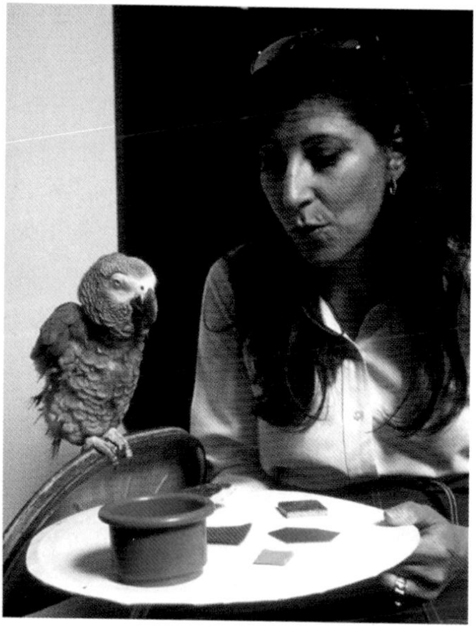

Alex and Irene Pepperberg whose thirty years' work together fundamentally changed the perception that parrots act out of instinct rather than intelligence.

A golden eagle during a training exercise. Its instinct to attack and destroy a hostile drone seems inbuilt.

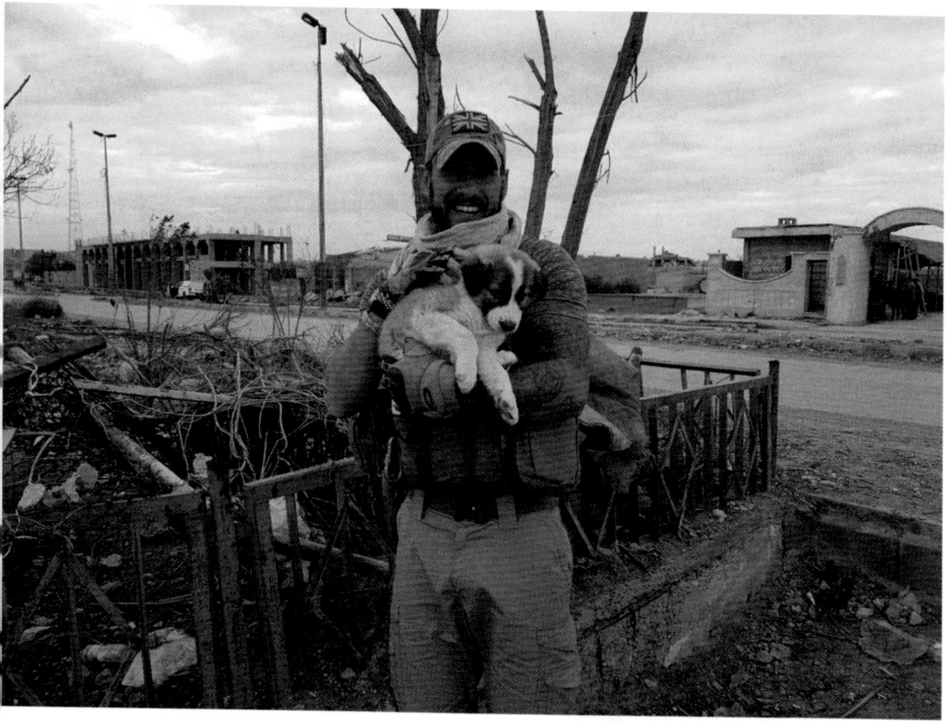

Sean Laidlaw with Barrie. The puppy he rescued in war-torn Syria went on to become his own saviour.

James Bowen and Bob, his lifesaver. Several best-selling books and a feature film turne Bob from street cat to celebrity.

Alexander the Great impresses his father and courtiers by taming the wild stallion Bucephalus. They became inseparable and together they conquered Asia.

Charisma, barely bigger than a pony, and Mark Todd, six foot three, made a magical partnership. Here they celebrate winning the gold medal at the Los Angeles Olympics in 1984.

:hristian with John Rendall (left) and Anthony Bourke who bought him as a cub
om Harrods in 1969 and later released him into the Kora Nature Reserve, Kenya.

Claire Guest and Daisy, whose scenting ability saved her life. Their work together pioneered a revolutionary way to detect cancer.

Diesel, the French police force's first dog to be killed in active service during a terrorist attack in Paris in 2015. She was posthumously awarded the Dickin Medal for her bravery.

Dolly the sheep, the world's first cloned mammal, meets the media in February 1997.

PC Dave Wardell and police dog Finn on the day he was awarded the PDSA Gold Medal 'for life-saving devotion to duty, despite being grievously injured while preventing a violent criminal from evading arrest'.

The sacred geese kept as guardians of the Temple of Juno. Their honking and flapping warned the Romans of the attack by the Gauls.

Mary Flood with Jake, who searched the rubble for survivors of the 9/11 attack on the Twin Towers in New York. He was also one of several rescue dogs who worked in the aftermath of Hurricane Katrina.

Hearing Dogs for Deaf People sent Jovi to Graham Sage, and his life was transformed. Jovi helped Graham pursue his passions for teaching and sport.

When Amit Patel suddenly lost his sight, he thought he would have to give up everything he held dear. Kika, his extraordinary guide dog, restored his freedom and confidence.

A postage stamp commemorating Laika, the first animal to be launched into space in November 1957.

A 1785 caricature of the Learned Pig, who wowed audiences across the country with his intelligence and skills.

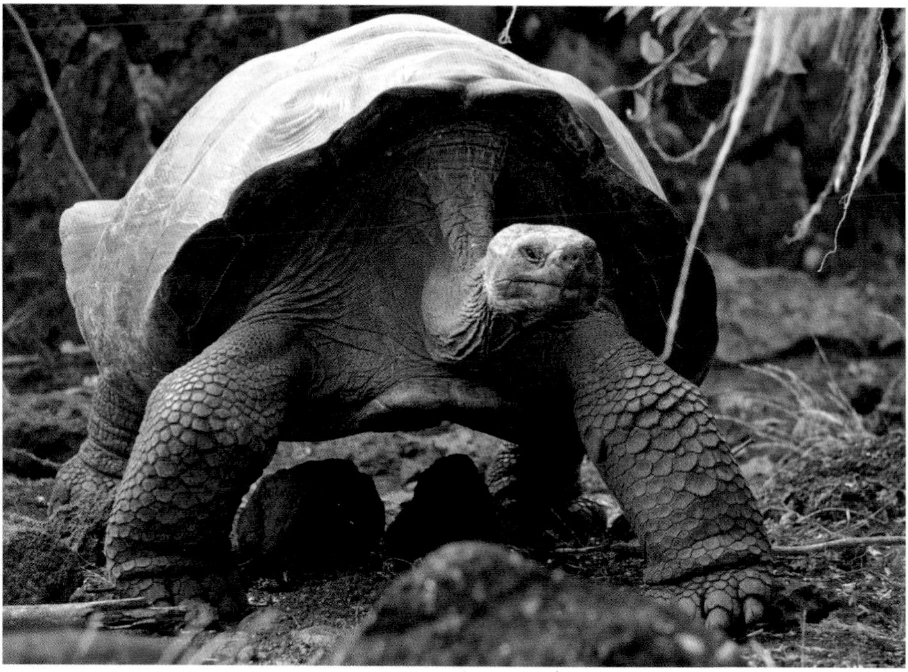

Lonesome George, a giant tortoise from the Galapagos Islands. Known as the rarest animal on earth, he became a conservation icon. When he died in 2012, the Pinta Island tortoise became extinct.

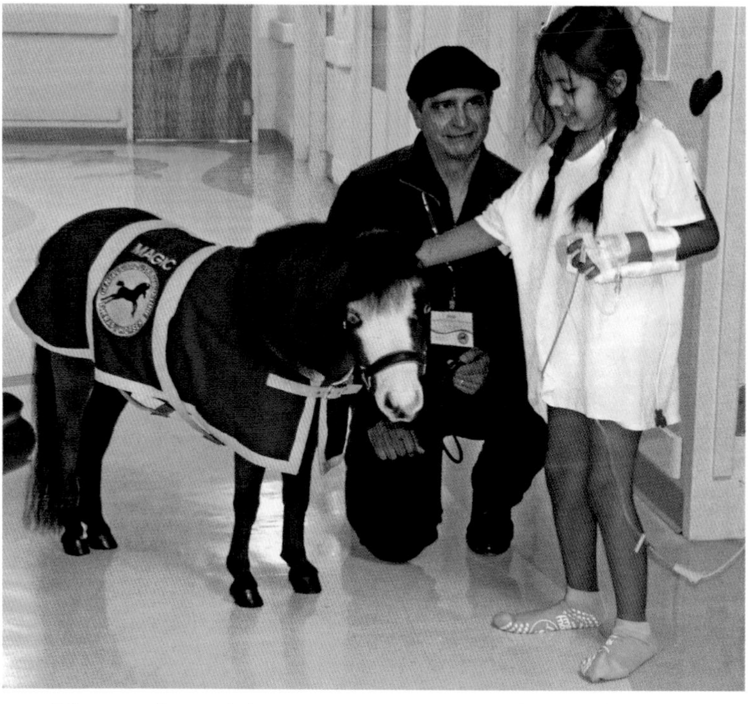

Therapy horse Magic, twenty-seven inches tall, visits a children's hospital in Chattanooga, Tennessee.

John Whitaker and Milton became the top show-jumping pair in the world. Here they are competing at Aachen in 1991.

Moko the Bottlenose dolphin plays with swimmers off Mahia beach, New Zealand. Succeeding where humans failed, he led two pygmy sperm whales trapped between a sandbar and the beach to the safety of the open sea.

behaviour, the ravens seem to like to display emotion. They have been known to mourn for their dead crow companions and they silently gathered around the Tower chapel after the death of one of the chaplains.

I still think they're spooky.

GUSTAV

Did you know that the animal that comes in at number two on the war hero list (behind dogs) is the pigeon? No less than thirty-two pigeons have been awarded the highest honour for animal bravery, the Dickin Medal.

During the Second World War, more than 200,000 pigeons were used – by the army, the RAF, the Civil Defence and the news services. Their contribution was so valued that they were given royal protection. There was a special Air Ministry Pigeon Section and a Pigeon Policy Committee which made decisions about the use of the birds in a military context. An RAF squadron was used to cull birds of prey along the coastline to prevent the pigeons being attacked on their missions, and anyone found to 'wound or molest' one faced a fine of £100 or a prison sentence.

Many birds were dropped behind enemy lines and would return carrying critical information from secret agents. Others were part of an intelligence-gathering operation codenamed Columba, in which questionnaires were dropped to civilians living under Nazi rule in Europe. The birds would return with the completed papers.

The National Pigeon Service loft on Thorney Island, West

Sussex, trained birds to take part in a number of important secret missions. Gustav (official name NPS.42.31066) was one of them.

Gustav had been donated to the cause as an eight-week-old 'squeaker' by a local pigeon fancier named Fred Jackson. He was trained in the same way as any other homing bird: allowed to fly around their own area then gradually taken further and further away. The birds use the position and angle of the sun to help them determine direction, as well as familiar landmarks. Research by an American geophysicist in 2013 also suggested that they use 'infrasound' (ultra-low frequency sound waves) that can't be picked up by humans to guide them home.

The bird was initially used to carry messages from the Resistance in occupied Belgium and proved to be a reliable courier. This marked him out as a candidate for other activity and he was given to Montague Taylor, a Reuters news reporter. Pigeons were used by war correspondents in much the same way as the military – to deliver important messages as quickly as possible – and Gustav soon became soldier and journalist rolled into one.

During the D-Day landings on 6 June 1944, Taylor was aboard a tank landing ship and selected Gustav for his most crucial mission of all. The reporter released him from his wicker basket and sent him home with the first news of what was happening on the Normandy shore. The famous message read: 'We are just 20 miles or so off the beaches. First assault troops landed 0750. Signal says no interference from enemy gunfire on beach . . . Steaming steadily in formation. Lightnings, Typhoons, Fortresses crossing since 0545. No enemy aircraft seen.'

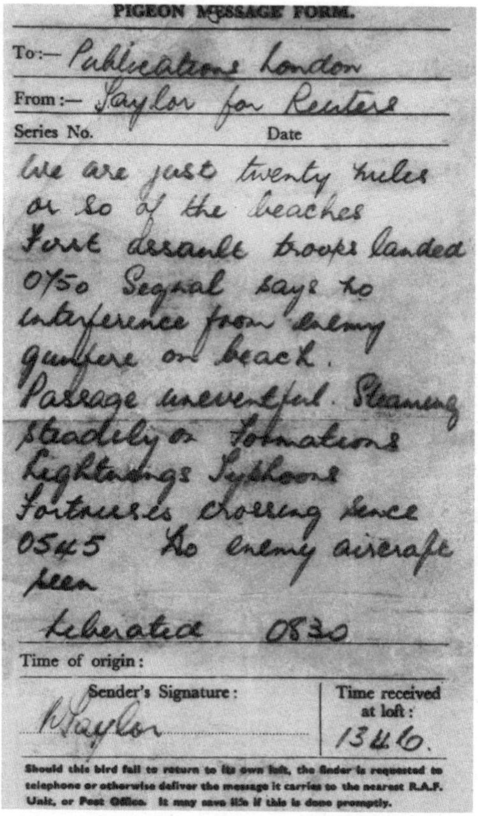

PIGEON MESSAGE FORM.

To:— Publications London

From:— Taylor for Reuters

Series No. Date

We are just twenty miles
or so off the beaches
First assault troops landed
0750 Signal says no
interference from enemy
gunfire on beach.
Passage uneventful. Steaming
steadily in formations
Lightnings Typhoons
Fortresses crossing since
0545 No enemy aircraft
seen

Liberated 0830

Time of origin:

Sender's Signature: M Taylor

Time received at loft: 1346.

Should this bird fail to return to its own loft, the finder is requested to telephone or otherwise deliver the message it carries to the nearest R.A.F. Unit, or Post Office. It may save life if this is done promptly.

Montague Taylor's message of 6 June 1944 carried by Gustav, informing that the Normandy landings had begun.

Gustav duly set off with the war correspondent's despatch on his journey of over 150 miles, into headwinds of thirty miles per hour. He avoided the hawks who had been trained by the Germans to attack pigeons on the coast. Five hours and sixteen minutes later, he arrived back at his loft on Thorney Island with word that the operation to end the Second World War had finally begun. His fellow messenger pigeon, Paddy, later delivered news that the D-Day landings had been a success.

Both pigeons were awarded the Dickin Medal and received a warm kiss from Esther Alexander, the wife of the First Lord of the Admiralty.

HACHIKO

One of the central tenets of Japanese culture is loyalty. It is regarded as the supreme virtue. Before the Second World War, every child was brought up to pledge loyalty to their country and their emperor. These days, loyalty to one's company and colleagues is paramount.

Perhaps the greatest Japanese story of devotion and fidelity of all time belongs to a dog. His name was Hachiko (ハチ公). He was a Japanese Akita, the much-loved pet of Hidesaburo Ueno, a professor at Tokyo Imperial University.

Akitas are working dogs, originating in the mountains of northern Japan. They look like little teddy bears as puppies, with thick, soft coats. They grow up to be strong and powerful and are excellent hunters. They were used as guards for high-profile royals or nobles in ancient feudal Japan. The Japanese government declared the breed 'a national monument' in 1931, and these days they are often used for police work. They are renowned for being fearless and faithful – none more so than Hachiko, the golden-brown Akita who became the entire nation's symbol of loyalty and fidelity.

Hachiko, or Hachi, was born in November 1923 and brought to live in Shibuya, Tokyo, by his owner Professor Ueno, who worked in the agriculture department at the Tokyo Imperial University. Every day, Ueno would walk to the station with Hachi and then commute to his office at the university. Every afternoon at 3pm they would meet at Shibuya Station again to walk home.

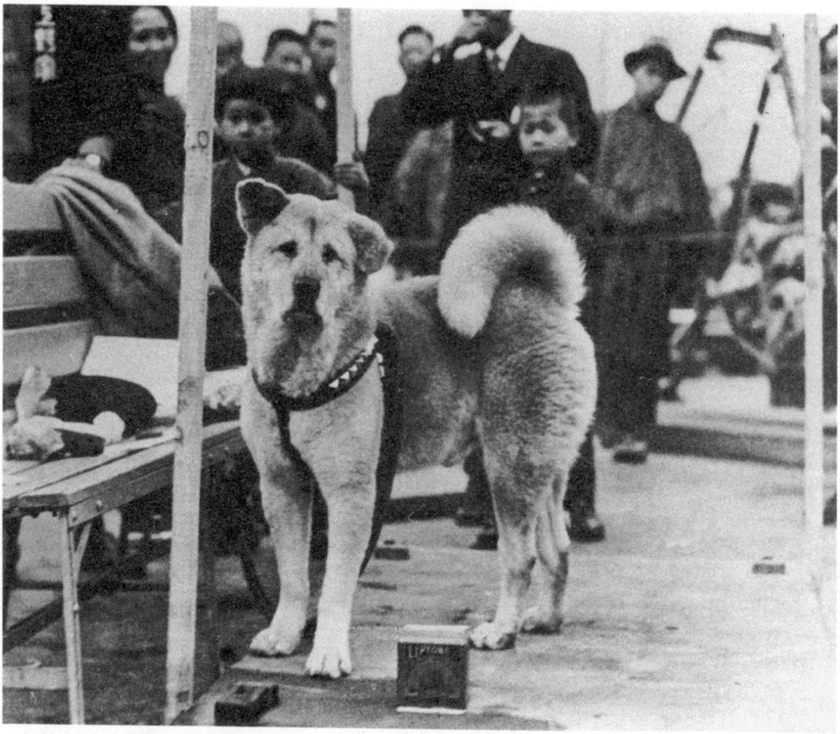

Hachiko waits for his master at Shibuya Station.

That system worked like clockwork until 21 May 1925 when the professor suffered a brain haemorrhage at work and died. Every day for the next nine years, at 3pm on the dot, Hachi waited at the station, hoping his beloved professor would turn up. It didn't take long for the dog to attract the attention of other people at the station, but none of them knew that he was waiting for a master who would never come. Then one of Ueno's students, Hirokichi Saito, spotted the dog at Shibuya and decided to follow him home.

Saito, who had a keen interest in Akitas, discovered that Hachi was now living with Ueno's former gardener. When he found out why the dog went to the station every day at 3pm, he wrote about it in a number of articles. The first of these was published in 1932 and put Hachiko firmly in the national spotlight.

The depth of the dog's loyalty to his master's memory touched people across the country. Some brought him treats and titbits to sustain him during his wait at Shibuya. Teachers used his vigil to set an example to their pupils.

Hachiko died in March 1935 at the age of eleven. A nation mourned. He was cremated and his ashes were buried alongside those of his beloved master in Aoyama Cemetery, Minato, Tokyo. His fur was preserved to be stuffed and displayed at Japan's National Science Museum.

Even before his death, in 1934 a statue was erected in his honour outside Shibuya Station, although it was melted down to help the Japanese war effort and then replaced in 1948. The entrance nearest to the statue is named Hachiko-guchi, or the Hachiko entrance/exit. There is another statue of Hachi outside the Akita Dog Museum in Odate, his original home town.

In March 2015, a further statue – a bronze of Hachi greeting his master – was put up outside the Faculty of Agriculture of the University of Tokyo to commemorate the eightieth anniversary of Hachiko's death. His memory is marked annually with a ceremony at Shibuya Station to remember the dog's unwavering devotion that became an example to all of Japan.

HAM

This story should come with a health warning because it really doesn't reflect well on us humans. We can be abhorrent at times, especially when exploiting animals in the name of science. I think it is important to recognise and salute the contribution of those animals who had no choice in the matter, yet played a part in our so-called 'progress'. One such is Ham the chimpanzee. Ham was named after the Holloman Aerospace Medical Center, the New Mexico lab that prepared him for his flight; his name was also inspired by the laboratory's commander, Lieutenant Colonel Hamilton 'Ham' Blackshear.

Ham, born in 1957, was captured in Cameroon as a baby. He was bought for $457 by the US Air Force in 1959 and taken to the Holloman Air Force base as one of forty chimpanzee candidates. At that point Ham was known only as No. 65, supposedly because officials were keen to avoid the bad press that would follow if a named chimp were to die in a failed mission. The odds were definitely against survival.

His training programme included various simple tasks in response to light and sounds. Later he was taught to push a lever within seconds of seeing a flashing blue light. When he got it right, he earned a snack. When he failed, he would receive a light electric shock on the soles of his feet.

Ham passed all his tests and was selected for a voyage into space on 31 January 1961, when he was launched on a sub-orbital flight from Cape Canaveral, Florida. He wore a nappy, waterproof pants and a specially designed space suit. He was strapped into a capsule that sat inside the nose cone of the Mercury-Redstone 2 rocket.

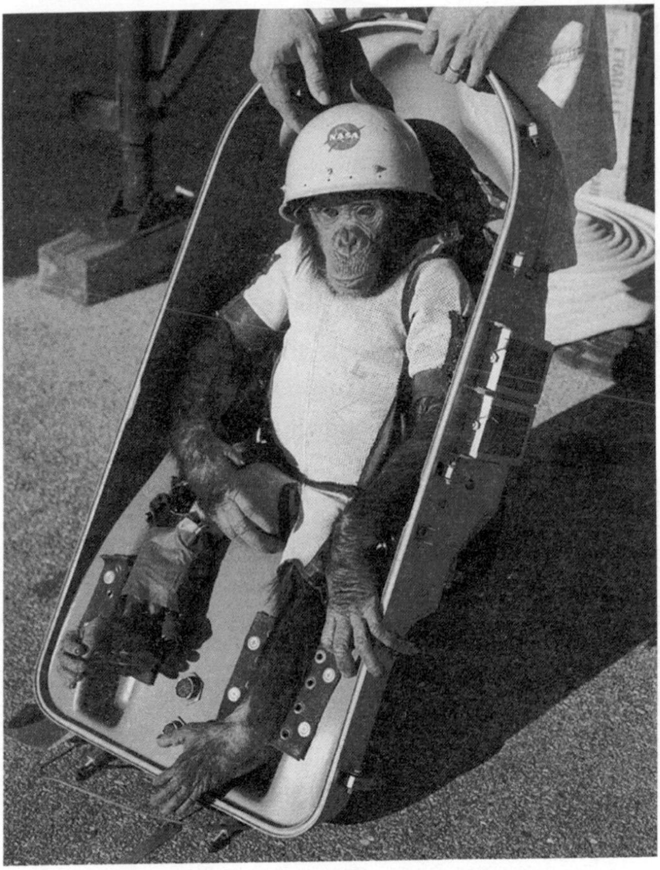

Ham at Cape Canaveral on his way into space, 1961.

His heart rate and breathing were monitored throughout the flight, during which he did his lever-pushing only fractionally slower than he had done in training. Sadly and cruelly, a malfunction in the valve meant he received a shock whether he got it right or wrong.

His lever-pushing proved to the lab's scientists that tasks could be completed during space flight and allowed them to check 'the craft's environmental control and recovery systems' as well as providing 'a first test of the functioning of the life support system during an appreciable period of zero gravity'. This historic experiment led directly to Alan Shepherd's test flight aboard Freedom 7 three months later.

The test flight on an unwitting and unwilling animal was widely criticised. Ham splashed down in the Atlantic after a 155-mile flight lasting sixteen minutes and thirty-nine seconds, emerging from the capsule fit and well, but with a bruised nose. However his trainer said of the recovery, 'I have never seen such terror on a chimp's face.' This sentiment was echoed by biologist Jane Goodall, undertaking a research project on chimpanzees in Tanzania, who saw raw footage of the flight and said Ham's face 'showed the most extreme fear'.

Ham spent the next seventeen years alone, the only chimp at the National Zoological Park in Washington DC. He was later moved to join others at North Carolina Zoo where he died in 1983, aged twenty-five (the average lifespan of a chimpanzee in captivity is thirty-one years).

The Armed Forces Institute of Pathology performed an autopsy, but plans for Ham to be stuffed and put on display at the Air and Space Museum in Washington were shelved after a public outcry. A leader in the *Washington Post* said: 'Talk about death without dignity. Talk about dreadful precedents – it should be enough to make any space veteran more than a little nervous about how he is going to be treated in the posthumous by and by.' Instead, Ham was buried at the International Space Hall of Fame in New Mexico, although his skeleton was kept for its 'scientific value'. Not the most distinguished of ends for the poor animal who helped achieve successful and safe space flight for humankind.

*

In November 1961, NASA sent a second chimp into space, this time to orbit Earth. Before his flight aboard Mercury-Atlas 5, Enos completed more than 1,000 hours of intensive training at the Holloman Air Force Base and at the University of Kentucky. On his flight he experienced weightlessness and G force in the same way as human astronauts Yuri Gagarin and Gherman Titov before him.

As with Ham, a malfunction exposed the chimp to far more electric shocks than intended (seventy-six in total), and this and other technical problems meant the operation was aborted before its planned third orbit.

Again, a preliminary flight by a primate allowed his human counterparts to complete a similar mission. John Glenn became the first American to orbit Earth in February 1962.

Enos died of dysentery in November 1962, closely observed by scientists who concluded that this was not related to his space adventures.

HOOVER

In May 1971, a Harbour seal pup was discovered on the shore of Cundy's Harbor, Maine. He was found by Scottie Dunning who, together with his brother-in-law George Swallow, set about searching for the tiny pup's mother. They found her lifeless body hidden among rocks along the shore.

The men were nervous about leaving the orphaned seal alone to meet a similar fate. After weighing up the options, Swallow, who was a local fisherman, decided to take the pup home. He kept him in the bath for a few days, until his family objected. Then he moved him to the garden to a spring-fed pond, complete with a specially constructed shelter in which the pup could sleep.

Swallow tried feeding the seal pup milk from a baby's bottle but he wasn't keen. A neighbour suggested trying ground fish and the youngster sucked that up enthusiastically – 'like a vacuum cleaner', observed an onlooker. So it was he earned the name Hoover.

Hoover became a much-loved family pet, the bond between him and his new family growing stronger by the day. When they took him on trips into town, he stuck his head out of the car window for all to see, just like a dog.

Swallow and his wife Alice talked constantly to the seal as they went about their daily chores, and Hoover would honk or bark in response. Then a miracle seemed to happen. When the seal was two months old, a group of children who had been playing with him at the pond, rushed to tell his owners that Hoover could talk.

The Swallows remained sceptical until the day that George, arriving at the pond, was greeted by Hoover with a gruff and hearty 'hello they-ah'. He couldn't believe his ears, especially as Hoover sounded exactly as he did, New England accent and all.

It seemed that Hoover had spent so much time with George and Alice that he'd not only learnt to speak, but understood what phrases he should use and when. He also loved to play hide-and-seek with his owners; and when George called out, the seal would waddle over to give him a wet and rather fishy kiss.

As the seal grew bigger, his appetite increased and soon the Swallows knew that neither their pond nor their supply of fish caught by George was big enough, and they couldn't afford the cost of buying fish from a local firm. Reluctantly they offered their beloved pet to the New England Aquarium.

George Swallow and Hoover in Boston, 1971.

As Swallow left Hoover at his new Boston home, he turned back and told the staff there, 'By the way, he talks.' They nodded, but it was clear that they didn't believe him. Indeed, the seal stayed silent until he had fully settled in to his new home which coincided with his starting to notice his female seal companions. Male seals normally sing to attract a mate but Hoover, having lived most of his life with humans, was not aware of this. So instead, his seduction technique consisted of him shouting 'whaddaya doing?' and 'get oveh heah' – still sounding uncannily like George Swallow.

Staff at the aquarium had never heard the like and employed a scientist to study Hoover's strange speech. At first scientists disagreed about whether or not Hoover had the ability to speak. It was not until they interviewed George Swallow and heard his heavy Maine accent that they realised the full nature of Hoover's amazing talent.

Not surprisingly, word of the loquacious pinniped began to spread, and people flocked to see the talking seal for themselves. Hoover was featured in the *New Yorker* and in *Reader's Digest*. He appeared alongside the Swallows on *Good Morning America*. His catchphrases, which also included 'hello there', 'how are you?' and 'come over here', delighted millions.

Scientists were baffled as to how Hoover had learnt to reproduce human speech quite so clearly. It was only in 2019 that research from biologists at the University of St Andrews proved that a number of factors (including sociability and intelligence) enable seals to imitate and learn many of the sounds that form human speech. A crucial aspect of the seal's ability is the similarity of their sound production mechanisms to those found in humans: they use the same structures in the larynx as we do to create sound – and to 'sing'.

In a long and painstaking process, the St Andrews scientists trained three seals, first to copy sounds from a regular pinniped repertoire, and then to copy new sounds by changing their formants, the characteristic pitch constituents used in human speech. In the end, all three were able to copy sequences of sounds, and one, Zola, to repeat melodies, including the first ten notes of 'Twinkle Twinkle Little Star' and part of the theme of *Star Wars*.

Vincent Janik of the Scottish Oceans Institute at the university explained that the study 'gives us a better understanding of the evolution of vocal learning, a skill that is crucial for human language development'. This research also shows that Hoover's natural abilities truly made him a phenomenon, perhaps one that will never be repeated. A visitor to the aquarium said, 'That little seal could swear worse than any sailor. He would mimic a lot of human speech . . . He was like a floating puppy dog with a dirty mouth.'

None of Hoover's six pups exhibited any signs of verbosity but his grandson Chacoda or Chuck is showing promise and signs of following in Hoover's footsteps, so listen out for any news.

To this day, Hoover remains the only seal to have had an obituary in the *Boston Globe*, and it's a mark of the impact he made on the Swallows that, on George's gravestone, alongside his own portrait is one of Hoover the talking seal.

HUBERTA

In life she was known as Hubert. It was only in death that it became apparent that the Zulu folk hero of the 1920s was actually female – and so history will for ever more record her as Huberta.

No one is quite sure what prompted the hippopotamus's 1,000-mile journey south from her waterhole in the St Lucia estuary in Zululand to the Eastern Cape, but her epic trek drew crowds wherever she went.

Huberta's journey began in November 1928. The *Natal Mercury* reported that a hippo had appeared in the area, feasting on sugar cane in a field. The article carried the only known photo taken of her during her life.

Huberta's next stop was about nine miles north of Durban, near the north-coast railway line at the mouth of the Mhlanga

River. There she managed to outsmart an attempt to capture her and put her in Johannesburg Zoo. She continued her journey, heading south along the coast of the Indian Ocean.

Hippos are known across Africa as one of its most dangerous animals. They are extremely protective of their young, fiercely territorial and, although they only eat grass, if a human being got in the wrong place at the wrong time, a hippo's enormous jaw would snap that person clean in half.

Consequently, Huberta's entrance onto the veranda at the Durban Country Club caused something of a stir. She appeared during an April Fools' Day party, so at first no one was sure whether this was some kind of elaborate joke. Either the cocktails were not to her liking or the life of a socialite didn't appeal, but Huberta left via the golf course and was later spotted in the doorway of a city-centre chemist.

Any celebrity can tell tales of the dark side of fame and, had she been able to articulate events surrounding her trek, Huberta would have been no different. There were fans and journalists but far more sinister were the hunters following her progress, desperate to bag what had become the best-known animal in South Africa.

She travelled mainly at night and became an expert at avoiding the hordes. Nonetheless, she was revered by many groups along the way. Local Indians serenaded her with drums and sacrificed a goat in her honour. The Mpondo believed that she was the reincarnation of a local witch doctor; the Zulus that she was connected to the Great King Shaka. In 1931, Natal Provincial Council declared that she was 'royal game', meaning that she was protected by law.

Most hippos will travel on land for six or seven miles to find food. Huberta's epic journey covered 1,000 miles and lasted three years. It took her through towns and cities, across railway lines, roads, 122 rivers, and through gardens and fields. Whether she was following in the footsteps of her ancestors, fleeing tragedy, looking for a lost mate or simply after a bit of adventure, by the time she arrived in East London on South Africa's southeast coast in March 1931, she had become a national heroine.

The story of Huberta's journey by G. W. R. *Le Mare, published in 1931.*

Tragically, her story does not have a happy ending. Just a month after she reached her destination, Huberta was shot by three farmers as she swam in the Keiskamma River. They claimed that they were unaware of her protected status, but after a public outcry (it was even discussed in the South African parliament), they were arrested and each fined £25, the equivalent of £400–500 today.

Huberta's trek and her untimely, needless death touched people across the world. It featured in publications from *Punch* to the *Chicago Tribune*. Her body was sent to the world's leading taxidermist in London.

An estimated 20,000 people welcomed her home in 1932 and watched her take pride of place in the Durban Museum. Today fans can find her on show at the Amathole Museum in King William's Town in the Eastern Cape.

There are much stricter laws now in place to protect hippos but there are less than 150,000 of them left in the world. Huberta was the most extraordinary of them all and proved that a hippo can certainly be a heroine.

JACK

British nurse Edith Cavell famously saved the lives of many soldiers from both sides during the First World War and helped around 200 Allied soldiers escape from German-occupied Belgium. Behind every great woman it seems is a great animal, and in this case it was a dog called Jack.

Cavell trained at the London Hospital in Whitechapel before moving to Belgium to teach at a number of hospitals. She was visiting her mother in Norfolk when war was declared, but instead of staying at home she decided to go back to Brussels where her clinic was taken over by the Red Cross.

The Germans invaded and occupied Brussels in November 1914. Cavell started sheltering soldiers from the Allied forces, as well as Belgian civilians of an age to be called up to the front who she helped escape to the Netherlands.

Cavell owned two dogs, Don who died before the war began, and Jack, a sheepdog-sized mongrel. During the occupation, Jack became a decoy: his walks were used to provide cover for soldiers as they made their escape. But the dog was walked so much that it began to raise suspicion, and Cavell and Jack were betrayed by an informant. Cavell was arrested in August 1915, tried and charged with treason. She was executed by firing squad two months later.

Edith Cavell in her garden in Brussels with her dogs Don and Jack, lying right.

Jack remained in Cavell's thoughts until the day she died. While in prison awaiting trial, she asked after him at every opportunity, instructing the sister from the hospital that he needed to be brushed daily. Jack's welfare was a priority: her notebook details her thoughts on everything from his exercise to diet and grooming.

Jack missed Cavell just as much and took her death very badly. He became bad tempered, biting the nurses and howling for his lost mistress. After being passed from pillar to post for the rest of the war, he was adopted by Princess Marie de Croÿ, who had worked with Edith to help Allied soldiers escape the Germans. He spent the rest of his life on her country estate and died there in 1923. His body was embalmed and sent to the Norfolk branch of the Red Cross, before a home was found for it with Cavell's notebook at the Imperial War Museum, where Jack remained on display until 2013.

JACKIE AND JACK

Another animal hero of the First World War was Jackie, a Chacma baboon who served in the trenches with the 3rd South African Infantry Regiment.

Baboons are the largest of the monkeys and the most distinctive with their bright red bottoms, overhanging foreheads and prominent long noses and jaws. Jackie was an unusual baboon in many ways, not just in that he preferred to wear uniform rather than expose his bottom to the world.

His military career began by accident, when Private Albert Marr joined up in 1915 and refused to leave his pet behind. Fortunately, Jackie took to army life like a duck to water, taking part in drills just like any other soldier in the regiment.

His comrades made him their official mascot. He was given his own rations, which he ate with a knife and fork, and he would light cigarettes for the other soldiers. Decked out in full uniform, Jackie was taught to salute the senior officers.

He took on shifts as a night watchman, his heightened sense of hearing making him a perfect candidate for the job.

It was all going swimmingly until the 3rd South African Infantry Regiment left for the front lines. Private Jackie went with them. In Egypt, Private Marr took a bullet to the shoulder and Jackie stayed by his side during the fighting, licking the wound to try and ease the pain. Then they were sent to the Western Front where Jackie kept Marr company on night watch. The baboon's ability to give early warnings of any impending attack was indispensable.

In April 1918, terrified by enemy shelling in Belgium, Jackie tried to build a stone wall around himself for protection. Under heavy fire, he was hit in the arm and the leg by shrapnel from an enemy bomb.

After the war, Jackie and Marr raised money for other injured troops by working with the Red Cross. Jackie was promoted to corporal, the only serving baboon ever to make it above the rank of private. He was discharged from the army with a military pension and his dedication to the war effort recognised by the award of a Pretoria Citizens Service Medal.

Jackie died tragically in a house fire in 1921, just a year after he and Marr had returned to South Africa.

Jackie wasn't the only South African baboon to make a name for himself in an unusual occupation (for a primate).

Back in the late 1870s, James 'Jumper' Wide was an employee of the Cape Town Port Authority Railway Service, his nickname earned from his habit of jumping between railway carriages, even when they were in motion. It seems obvious that this was an accident waiting to happen, and so it was that one day Jumper misjudged the distance and fell beneath the moving train, which severed both his legs at the knee.

Jumper was briefly bowed, but certainly not beaten. He fashioned two wooden legs out of pegs and used a trolley to help himself move around Uitenhage Station, where he had taken a new job. But working life became a struggle. He needed an assistant.

James 'Jumper' Wide and his assistant Jack.

The baboon he came across at a local market wasn't necessarily the most obvious candidate. Still, Jack, as he called his new helper, shaped up pretty quickly, learning to switch the train signals, hand keys to conductors and even to wheel Jumper round the station. Word soon spread about the simian signalman and he often had an audience, though it never put him off.

It is, perhaps, understandable that some people were a little unnerved at the idea that a baboon could be controlling their safety on the railways. And when the authorities discovered that the station assistant was actually a baboon, a manager was despatched to fire the pair forthwith.

But Jumper was persuasive and the manager agreed to test the baboon's skills. Jack kept his job, becoming an official employee of the railway and earning twenty cents a day and half a bottle of beer a week. He continued to work the signals for nine years and never made a single mistake.

JAKE

It is amazing how many of our heroic animals had unfortunate starts in life. Jake's story proves that however badly humans may have behaved, animals will still come through for us. Their bravery, loyalty and kindness should teach us all how to be better people. I think in many ways the COVID-19 lockdown reminded us that the simple things in life are the most important – getting outside for a walk, spending time concentrating on domestic tasks such as cooking, cleaning or gardening, looking out for each other and not rushing from one adventure to the next. We became like dogs in our reliance on four basic things: food, sleep, exercise and love.

Jake the black Labrador had none of those things when he was abandoned as a puppy, left to wander the streets with a broken leg and a dislocated hip. He was adopted at ten months old and nursed back to health by a member of Utah Task Force 1, a federal search-and-rescue team based in Salt Lake City.

When Mary Flood looked at her new puppy, she saw more than just an injured stray. She saw that Jake had a strong survival instinct and a determination to succeed. Also, she knew he was brave and had seen enough in his young life to know that he could face any danger.

Once he was fully fit, Mary trained Jake to become part of an elite force of search-and-recovery dogs. These are specialist dogs who are sent into disaster areas within the first twenty-four hours to find survivors.

Jake's toughest mission came in 2001, as he dug through the wreckage of the World Trade Center after the 9/11 attacks in New York City. He worked at Ground Zero for more than two weeks, one of 300 dogs and part of a 10,000-strong rescue team engaged in the search for human life.

The task was brutal for all involved. Jake and the other dogs worked twelve-hour shifts, searching the razor-sharp rubble in near-impossible conditions. Vets stepped in regularly to clean the dogs' eyes, noses and paws before they were sent back in again to do their job.

The last person to be pulled from the wreckage alive, twenty-seven hours after the towers fell, was Genelle Guzman-McMillan. After that, rescue turned into recovery as the dogs retrieved bodies of those who had been killed. They found the body parts and belongings of people who might not otherwise have been identified at all. More than 2,600 people lost their lives when the Twin Towers fell.

The dogs involved in the search, rescue and recovery mission were heroes and were honoured as such. And when Jake walked into a smart Manhattan restaurant wearing his search-and-rescue vest, he was treated to a free steak dinner.

Jake was 'always ready to work' according to Mary Flood, which was lucky as he was on twenty-four-hour call, and summoned immediately if a flood, earthquake, avalanche, hurricane or building collapse occurred anywhere in the country.

In 2005 Mary and Jake drove the thirty hours from Utah to Mississippi, where Jake searched flooded homes for survivors after Hurricane Katrina. Later that year he travelled to New Mexico to help in the aftermath of Hurricane Rita, which killed 120 people in the area around the Gulf of Mexico. He never once hesitated to perform his work at the highest level, despite the difficult and dangerous conditions he faced.

Jake was also a good teacher and helped train the next generation of specialist rescue dogs. He taught them to follow a scent in all weather conditions, to detect a scent high up in a tree or building. Jake's calmness under pressure also made him a remarkable therapy dog. He worked with children at a camp for burns victims in Utah, as well as with adults in old-people's homes and hospitals. Mary said he was 'a great morale booster' even if he had a tendency to tidy up any unattended food left lying around.

When Jake died in 2007 at the age of twelve, his death was reported by news outlets across the world. There was speculation that his work at Ground Zero might have been behind his hemangiosarcoma, though scientists believe that there was no link between his heroic work on the site of the Twin Towers and the blood cancer that killed him. His ashes were scattered in the rivers and hills in Utah where he loved to swim and run.

Among the other canine heroes who helped rescue people on 9/11, the first dog into the wreckage was Appollo, who arrived just fifteen minutes after the towers collapsed. Falling debris and flames made his work impossibly dangerous, but he refused to let anything deter his efforts.

The last survivor of the World Trade Center attacks was discovered by Trackr, a German Shepherd. Afterwards he collapsed from smoke inhalation, burns and exhaustion and had to be treated with an IV drip. He died in 2009, at the age of fourteen, but lives on through the five puppies that were cloned from his DNA.

Sage, a Border Collie from New Mexico, searched for survivors of the attack on the Pentagon. She was only two years old and it was her first high-profile assignment. She spent a week searching for human remains, including those of the terrorist who had flown Flight 77 into the US Department of Defense headquarters in Arlington, Virginia. She was one of only fifty or so dogs to hold the highest level certification recognised by the Federal Emergency Management Agency (FEMA). Sage went on to serve in Iraq. She died in 2012, three years after winning the Search and Rescue category at the ACE Awards for Canine Excellence.

JET

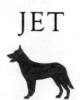

Jet, a pedigree German Shepherd from Liverpool, was born Jett of Iada in 1942, at the height of the Second World War. Loaned to the War Dogs School in Gloucester at the tender age of nine months by his owner, Mrs Babcock Cleaver, he was trained in anti-sabotage work that saw him spend eighteen months working on airfields. He then had further training in search and rescue.

Jet was assigned to handler Corporal Wardle and sent to London to become the first official rescue dog to work with the Civil Defence Service. His posting didn't start well as he was sick several times en route and never did become a fan of travelling in the back of a van or lorry, but he quickly proved his worth in the war-torn capital.

Jet was fearless, to the point that sometimes Wardle had to hold him back from entering a burning building until the ferocity of the flames had been dampened down. He was also determined, refusing to give up hope of finding survivors long after others had stopped searching.

Jet searching ruins for victims of a bombed hospital.

On the site of a bombed hotel in Chelsea, the rubble had been gone through a number of times and rescuers concluded that no one could possibly still be alive. But no one could persuade Jet to leave the collapsed building. The dog concentrated his efforts on an unstable section of wall that remained intact, refusing to move for twelve hours, until finally one of the men clambered cautiously to the top and, high up on a ledge, discovered an elderly woman, trapped but alive. Somehow Jet had known where she was, although she was too high above ground for most dogs to have picked up her scent.

Historian Ian Kikuchi, a curator at the Imperial War Museum in London, remarks that 'Even when searching piles of the remains of factories full of dangerous chemicals and poisonous smoke, [Jet's] incredible sense of smell was still able to detect survivors.'

In the course of his work, Jet recovered a total of 150 people and was awarded a Dickin Medal in January 1945 'for being responsible for the rescue of persons trapped under blitzed buildings while serving with the Civil Defence Services of London'.

Jet went back home to Liverpool after the war, but his heroic exploits were not over. In 1947 there was an explosion at a pit near Whitehaven, Cumbria. Jet's search for survivors there earned him the RSPCA's Medallion of Valour.

Jet died in 1949 and a memorial to his bravery stands in the flower garden at Calderstones Park in Liverpool.

JIM

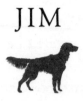

There are many ways that our four-legged friends can amaze and inspire. They can detect all sorts of things by smell, their hearing is far better than ours and, in some breeds, their sight is exceptional. However, it's unusual – unique even – to be able to identify a car by its colour, make and registration number, isn't it? It's also pretty clever to be able to predict the winner of the Kentucky Derby seven years in a row. Then again, Jim the wonder dog was more than a bit special.

If Jim were alive now, he'd be drafted in by *Top Gear* as a co-presenter and would work at weekends as a tipster on Sky Sports Racing.

Jim was a black-and-white English Setter (or specifically a Llewellyn Setter) from Louisiana, USA, born in 1925. He was considered the runt of the litter and while his brothers and sisters sold for $25 (around £300 in today's terms), he was valued at half that. Sam Van Arsdale of Marshall, Missouri, got a bargain.

It didn't seem that way to begin with, as Jim would simply lie in the shade watching the other dogs being trained to work in the fields. Maybe he was too clever to exhaust himself running around and could learn more by observing. He soon proved to be a top-class hunter, helping his master bag over 5,000 birds (before he stopped counting). *Outdoor Life* magazine described him as 'The Hunting Dog of the Century'.

Jim's other claim to fame was discovered quite by accident, when his owner suggested they go and shelter from the sun under a hickory tree. Although there were many other types of tree around them, Jim picked the correct one first time. It proved to be no fluke.

Van Arsdale put Jim's powers of understanding to the test in many different ways. His dog could go into the street and pick out a car by its make, colour, number plate. He could select people from a crowd simply when asked to find 'the man who takes care of sick people' or 'the one who sells hardware'.

He could carry out commands given in any language, including Morse code, and could seemingly read written commands. He predicted that the Yankees would win the World Series (baseball) of 1936 and picked seven Kentucky Derby winners in a row. He could even prophesy the sex of unborn babies.

The head vet at the University of Missouri, Dr A. J. Durant, could find nothing exceptional about Jim's physiology and other scientists were baffled. A group of students put the dog to the test and he passed every challenge. Durant concluded that Jim 'possessed an occult power that might never come again to a dog in many generations'.

It seemed Jim really was a 'wonder dog', the name given to him in a press review of a performance in Wyoming in 1935. He made it into the pages of *Ripley's Believe It Or Not!* and his fame spread far and wide.

Jim died, aged twelve, in 1937 and his grave is the most visited in Marshall's Ridge Park Cemetery, with tributes such as coins and flowers often left on the stone. Friends of Jim the Wonder Dog is an organisation dedicated to preserving the history of Marshall's most famous 'son' and has raised the money for a garden and statue in his honour, on the site of the Ruff Hotel where he lived.

JOVI

Now to a present-day hero. I have learnt a lot about the work of Hearing Dogs for Deaf People through my work at Crufts and I have even experienced a dog 'waking me up' from a mid-afternoon snooze (all in the name of TV, I should stress). It is less well known as a charity than Guide Dogs, perhaps because it was established some fifty years later in 1982. It has around 1,000 working partnerships in the UK, where the dog alerts their owner to sounds that they cannot hear themselves. As well as making sure they can respond to an alarm or a doorbell, for someone with limited or no hearing, a dog changes their whole outlook on life. Deafness can be very lonely and often leads someone to retreat from situations they can't join in. The right dog can change all of that.

Graham Sage began to lose his hearing at fifteen, gradually at first. By the time he started university, it was clear that things had become considerably worse and he was really struggling to keep up. When his lecture theatres proved too large for lip-reading he finally had to admit that he had a problem.

Diagnosed with Ménière's disease and tinnitus by a specialist, Graham's hearing still continued to deteriorate. By the age of twenty he had become profoundly deaf. He found the thought of wearing hearing aids embarrassing but knew that if he couldn't hear a smoke alarm, for example, he might be in serious danger.

Daily life became increasingly difficult. He couldn't hear

the doorbell. He was so concerned that he might miss his morning alarm that he was barely sleeping. His studies and social life began to suffer. It was clear that something had to be done.

Then Hearing Dogs for Deaf People sent him Jovi, and Graham's life was transformed. Once introverted and withdrawn in social situations, he began to enjoy being around other people: '[Jovi] has helped me to overcome some of my anxieties around interacting with other people, and his presence actually encourages interaction', Graham explains.

> I have had so many people come up to me and ask questions about him. When people find out Jovi is my hearing dog they start to ask more questions. This has made me far more accepting and even proud of my hearing loss.

Jovi has helped Graham choose a career he loves. Working as a teacher might have seemed an impossibility for someone unable to hear children, school bells or security alerts, but thanks to Jovi, Graham has been able to follow his passion. He carefully alerts Graham when timers go off to signal the end of a lesson, and helps him to teach each new intake to be more deaf-aware and to communicate in a way that allows him to lip-read. And after seeing what Jovi did for Graham, the whole school was inspired to help raise £20,000 for Hearing Dogs. One colleague even ran the London Marathon dressed as a dog, taking the world record for fastest female to run the event in full animal costume.

Over the years, having a hearing dog has made Graham less reliant on other people, especially his wife Anna, giving him the confidence to think about starting a family: 'It is comforting to know that Jovi can be trained to alert me to a baby's cry and will add to the safety and security of the household.'

Jovi has even given Graham the confidence to enjoy his passion for sport. He captained the England Deaf Rugby Union side, going on to become their assistant coach. 'Jovi helps me to lead a "normal" life and I'm so thankful to him for that.'

JUDY

Few dogs can claim to have survived a prisoner-of-war camp, let alone kept up the morale of the inmates. But Judy, a liver-and-white Pointer from Hong Kong, did exactly that, as well as surviving a catalogue of disasters during the course of the Second World War.

Her story begins in 1936, in the Far East, when Judy became ship's mascot of the gunboat HMS *Gnat*. Although pointers are bred to be gundogs, Judy didn't take to this line of work. When the sailors sent her to retrieve imaginary birds as part of her training, she repeatedly fell overboard and the boat had to keep wheeling around to retrieve her from the sea. She proved much more successful as a sentry, alerting the crew to the presence of river pirates and enemy aircraft.

In 1942, Judy was on board HMS *Grasshopper* in the South China Sea when Japanese aircraft targeted the ship. The damage was so bad that everyone had to abandon ship and swim to land.

Dog and sailors found themselves on a remote island, along with survivors of another ship, HMS *Dragonfly*. The landscape was hostile. There was no fresh water and the men feared that they would die of thirst, but Judy was the one to save them. One day they found she had dug a big hole and unearthed fresh water for them all. With Judy, the sailors then trekked hundreds of miles through the jungle for five weeks, trying in vain to catch the last evacuation ship leaving Padang, Sumatra.

After the Japanese invaded Sumatra, the sailors and their mascot were captured and taken to a prisoner-of-war camp in the north of the country. The prison rations were barely enough to survive on but Frank Williams, a twenty-three-year-old airman from Portsmouth, shared his food with Judy. It was the start of a profound friendship and the pair became inseparable.

Frank Williams and Judy, the only official canine prisoner of war, at the Dickin Medal ceremony in 1946.

Conditions in the camp were harsh and the camp guards were notoriously brutal. Judy often risked her life trying to protect her crew by distracting the wardens as they meted out punishments, or warning them that a guard was on his way.

When the prisoners were shipped to Singapore, the Japanese refused to allow the dog to go with them but there was no way the men were prepared to leave her behind. They taught Judy to hop into a rice sack and carried her on board. Pointers are large dogs; Judy had to remain upside down, silent and completely motionless for over three hours hanging over Frank's shoulder until they got onto the ship and he could finally put her down.

Judy's adventure was not yet over. A British submarine, unaware that the SS *Van Warwyck* was carrying Allied forces, unleashed their torpedoes. The ship was destroyed, killing around 500 of the 700 prisoners on board. Judy was pushed through a porthole and managed to run down the side of the ship and escape. Even then, she remained loyal to the men around her, pushing pieces of driftwood towards those who were struggling to swim, saving several lives.

The survivors – including Judy – made it onto a Japanese tanker where the guards were not best pleased to see a dog. They declared that she would be executed as soon as they reached land. Only an intervention from the commander of her previous camp saved her. On his orders, the dog was listed as an official prisoner of war, making her eligible for rations and offering her protection from further threats of execution. As 'Prisoner No.81A', she became the world's first ever animal prisoner of war.

The sailors spent the rest of the war labouring in the jungle, building what became known as the 'Death Railway' between

Burma and Thailand. Over 100,000 prisoners and slave labourers died laying down the tracks for it. Frank Williams later said it was Judy who gave him the reason to live: 'All I had to do was look at her and into those weary bloodshot eyes and I would ask myself: What would happen to her if I died?'

Judy and Frank survived all kinds of hardship (including attacks from wild dogs and an alligator) to make it back safely after the war. Word of Judy's actions had already reached British shores and she received a hero's welcome, as well as a Dickin Medal. Judy and Frank then spent a year travelling around the country offering comfort and solace to the relatives of those who had died working on the Death Railway with them. Judy and Frank stayed together until she died in 1960 in Tanzania. Frank made sure that all her many achievements were recorded on her memorial stone. She was an exceptionally brave and intelligent dog.

KIKA

Amit Patel was working as a trauma doctor when he went blind. Overnight his world changed. In his final year as a medical student at Cambridge, he had been diagnosed with an eye condition called keratoconus, a condition which changes the shape of the cornea. It does not usually result in blindness, but Patel's body kept rejecting cornea transplants and eighteen months later blood vessels burst at the back of his eyes.

He woke up one morning and found he could not see. He

had to give up his job in Accident and Emergency at a London hospital. He was in constant pain and describes the despair at losing his sight as 'unfathomable'. All he could think about was what he would no longer be able to do. His plans and dreams lay in tatters. Things got so bad that he tried to take his own life.

Although he had the unwavering support of his wife Seema, adjusting to life without sight felt impossible. It was hard to imagine brighter days ahead. He threw himself into the practical tasks such as learning to use a white cane and to read Braille, but refused to entertain the idea of having a guide dog. 'Why would I put my safety in the hands of an animal so closely related to a wolf?' he asked, and went on,

> With the cane, I was in control. A stick is never going to drag you across four lanes of traffic in pursuit of a squirrel, or be distracted by a doughnut. Was it really possible that I could ever trust an animal with my life?

He was also concerned that he and Seema would not be able to cope with another living being in the house:

> I could barely look after myself after losing my sight, so I thought it would be a struggle to look after a dog. I had no idea what I would need to do, and thought it would be a lot more difficult.

Perhaps Seema had other ideas because she started to volunteer at Guide Dogs for the Blind and suggested to Amit that he meet the Engagement Officer Dave Kent, who was also blind. He promised to look after them both and enrolled Amit on a four-month assessment.

It can take up to two years to find a suitable dog, but less than two months later Amit and Seema were asked if they wanted to meet a young dog with a strong personality. She was a Labrador called Kika and her issue was that she either really liked people or she didn't. At all.

Amit waited nervously. Kika was headstrong and there was a strong chance she might reject him. The early signs were positive – she didn't growl at him or walk out – but there was a long way to go.

Guide dogs are trained to stop at kerbs and steps, to walk in the middle of the pavement and avoid obstacles. They are taught to remember regular routes or features of favourite places and not to turn corners unless instructed. The new owner has to learn the specific cues the dog has been trained to respond to and practise them under supervision as they go through a number of real-life situations.

It's like dating, and both partners in the new relationship have to be happy for it to continue. They had a week at home which went well, although Amit was still nervous about putting his trust in a dog. He struggled to let Kika take charge, pulling her back time and again. He wasn't sure it was going to work.

The next part of the course was residential and they stayed together at a hotel. Amit woke early and felt his way to the bathroom but Kika blocked his way and nothing he could do or say would get her to budge. In desperation he finally rang one of the instructors who reminded him that he was in charge and he should pick her up if he had to. It was only when he finally got into the bathroom that he understood the reason for the dog's refusal to move: 'The floor was covered in water at least an inch deep and it was slippery. I realised that Kika had been blocking my way because she knew there was a hazard on the other side of the door.'

That moment changed everything. He finally understood that Kika only cared about his safety and was humbled by everything she did for him. He was overjoyed when they passed the course with flying colours and he was allowed to take her home. At long last he felt everything was going to be all right.

The first-known attempts to train guide dogs came in the late eighteenth century at a hospital in Paris, although there seems to be a picture of a dog leading a blind man on a mural in the Roman ruins at Herculaneum. However, it wasn't until the First World War, after soldiers started to return from the front blinded by shrapnel wounds or poisonous gas, that the idea of guide dogs as we know them today came about.

After observing how a dog behaved with one of his patients, Dr Gerhard Stalling opened the world's first training school in Oldenburg, Germany, in 1916. Over the next decade branches opened across the country, training up to 600 dogs a year to be used in Europe and America. The schools spawned other such training centres, for instance, L'Oeil qui Voit or the Seeing Eye in Germany, the US and Switzerland.

This inspired two British women, Muriel Crooke and Rosamund Bond, to train the UK's first four guide dogs in Wallasey, Merseyside, in 1931. Three years later the Guide Dogs for the Blind Association was founded. Now there are 4,700 dogs working with blind or partially sighted people across the UK.

While Patel had to give up his job in A&E, he now works as a diversity and inclusion consultant. On his daily commute

Kika helps him avoid hazards, including people she doesn't like the look of, and keeps him safe. She has given him freedom and confidence.

On their travels he directs her and she finds the best way to help him to his destination safely. If there are train changes, he asks her to find a member of staff to help. He explains, 'This is easy for her, as she recognises them by their high-vis jackets – though she has occasionally introduced me to groups of startled building contractors.'

When Amit and Seema's first child was born in 2016, Kika sniffed the baby tentatively before taking on the role of chief protector. Thanks to her presence, Amit can now take the pushchair out on his own:

> Thanks to one extraordinary dog I can live a wonderfully ordinary life. I can be a dad, a husband, a colleague, a friend and a neighbour. With Kika's help, I do it slightly differently. But we do it together and that's what it's all about.

KOKO

For nearly half a century, Koko the gorilla taught us about animal communication and the ability of some species to show a huge depth and range of emotion. She was intelligent, kind, creative, mischievous, cheeky and had a sense of humour. She also had a thing for cats.

In the wild, gorillas use a number of ways to connect with each other. These include posture, gesture, facial expression and more than twenty different sounds. Singing, for example, indicates contentment, while screams or roars denote anger. Mountain gorillas beat their chests, either to signify danger or to attract the opposite sex by showing off how strong they are. Others might greet each other with an embrace or by touching noses.

We have many things in common with primates including the need to form close bonds – in essence, an understanding of love or friendship. The development stages of our offspring are similar, but humans have larger brains, which in theory leads to greater intelligence and a more sophisticated use of complex language. In trying to work out what makes us human, it helps scientists to study the behaviour and communication tools of apes.

Koko was an exceptional subject. A Western lowland gorilla, she was born at the San Francisco Zoo in 1971, the fiftieth gorilla to be born in captivity. She spent most of her life with psychologist Francine 'Penny' Patterson and became an integral part of Patterson's research programme on primate communication.

Her training began at the age of one. Patterson spoke to her at every opportunity. Koko was reported to understand 2,000 words of spoken English, including concepts such as good and bad. In GSL, or gorilla sign language, she had an active vocabulary of more than 1,000 signs. Tests put her IQ somewhere between seventy and ninety (the average for a human is 100), but experts had different opinions about how closely Koko's progress aligned with that of human linguistic development. While some felt it mirrored that of a child, others believed that Koko did not understand meaning and learnt only on a reward basis.

Penny Patterson and Koko in California in 1972.

The way Koko used sign language was substantially beyond basic. She seemed to understand and to apply logic as well as emotional depth. Patterson reported that Koko even invented new signs, such as putting 'finger' and 'bracelet' together to mean ring. When Patterson published her research in 1978, critics suggested that Koko's actions were largely due to the Clever Hans effect – when an animal (or human) simply senses what someone wants them to do, and any signals may be unintentional.

In 1983, Koko asked for a cat for Christmas. Yes, you heard me. Not surprisingly, this caused quite a stir. In an interview with the *Los Angeles Times*, biologist Ron Cohn told reporters that she spurned the initial gift of a stuffed-cat toy, so on her birthday she was allowed to choose a real kitten from a litter that had been abandoned. She chose a grey male, a rolled-up bundle of fluff, and named him herself. She called him All Ball and cared for him lovingly, carrying him like a baby. Sadly, in December 1984, the kitten was run over by a car. When Patterson told Koko the news, the gorilla was visibly distraught. She signed the words for 'bad', 'sad' and 'cry', and then signed 'sleep cat'.

When she was forty-four, Koko picked out two kittens that she wanted to mother. The resulting behaviour, which you can watch on her YouTube channel Kokoflix, is adorable.

Koko was a media darling, even if her obsession with the word 'nipple' made any live interview something of a risk. Many celebrities came to meet her and she seemed to bond particularly with the actor Robin Williams.

After Patterson's research was complete, Koko went to live on a Gorilla Foundation reserve in Woodside, California, first with another signing gorilla named Michael, and then with another male, Ndume, until she died in her sleep aged forty-six in 2018. A statement released by the Gorilla Foundation said, 'The impact has been profound and what she has taught us about the emotional capacity of gorillas and their cognitive abilities will continue to shape the world.'

KUZNECHIK

Some years ago, I presented a programme on Radio 4 about racing camels. I went to Dubai to find out about the latest techniques for breeding from good racing cows (the females tend to be better than the males) without taking them off the racetrack. Essentially it came down to artificial insemination and embryo transfer, and I had to describe various elements of the process in detail for the audience. Perhaps it was too graphic in parts for a morning programme. I know I got complaints.

Anyway, my point is that a good racing camel is worth a lot. Camels have always had a high value because of their strength and stamina – they can carry 900 pounds for twenty-five miles a day and can turn on the speed if necessary. They are also renowned for being hardier than almost any other mammal. They can close their nostrils in a sandstorm; they can protect their eyes with their three sets of eyelids and two rows of eyelashes. They can go for weeks without water and then drink forty gallons all in one go, storing the excess in their humps. They can sit or kneel on hot sand because of their thick knee pads, and their thick lips mean they can eat even the thorniest of bushes.

They've also played their part in battle over the centuries. Camels have many advantages over the more traditionally used horses. They can carry a soldier plus all his supplies and equipment with ease, but still cover up to six miles an hour at a trot. They are bold, and far less likely to be spooked by gunfire and artillery than their equine counterparts. In 853 BC, the Arab king, Gindibu, sent 1,000 camels to fight in the Battle of Qarqar in Mesopotamia (modern Syria). In 1798, Napoleon formed a camel corps to fight in Egypt and Syria. In 1916, the British Empire pressed almost 5,000 camels into service in the Middle East. The resulting Imperial Camel Corps Brigade (ICCB) was made up of four battalions, two from Australia and one each from New Zealand and Great Britain. The brigade was part of the

Egyptian Expeditionary Force before being disbanded in 1919.

In the Second World War, the Soviet Red Army, hampered by a lack of auxiliary vehicles and the difficult terrain of the Kalmyk Steppes, used camels to transport everything from fuel and ammunition to wounded soldiers in the aftermath of the Battle of Stalingrad. One such animal was Kuznechik who has become a symbol of all the camels who have carried humans and their supplies in battle.

He was a Bactrian (two-humped) camel, conscripted in 1942 into the Soviet 308th Rifle Division – later renamed the 120th Guards Rifle Division – to carry food and cooking equipment. His height also came in handy for locating the division's camp from a distance. His name, Kuznechik (Кузнечик), means 'Grasshopper'. As the Rifle Division made its advance into Germany, fighting numerous battles along the way including the East Prussian Offensive (between January and April 1945), Kuznechik marched at the rear, carrying out his duties.

Of the 350 camels that took part in the war, many were killed in action. Others were left with zoos in Eastern Europe after being demobbed by their brigades. What happened to Kuznechik has rather divided opinion. Some argue that he was killed in an air raid near the Baltic Sea in 1945; others, more poetically, that he made it all the way to Berlin, whereupon his driver led him to the steps of the Reichstag to spit upon the ruined building.

Either way, Kuznechik's brave and loyal contribution to the Soviet war effort was recognised with a medal 'For the Defence of Stalingrad' and three wound stripes for the injuries he had sustained.

LAIKA

We have already heard tales of animals sent successfully into space to pave the way for manned operations to succeed. They all leave a bad taste and the story of the very first living creature to go into orbit is, I'm afraid, equally controversial and unhappy.

Laika (Russian for 'Barker') was a stray plucked from the Moscow streets who played a key part in the space race. From the humblest of beginnings, she managed to touch the world and spark a global debate on the ethics of using animals in scientific testing.

In 1957, the Soviet scientists successfully launched the first artificial earth satellite. They named it Sputnik 1. Their president, Nikita Khrushchev, was keen to build on this success as fast as possible. Sputnik 2 was rushed out within a month, a 'space spectacular' that would see a living being launched into space for the very first time.

Scientists believed that a stray who had survived the freezing Moscow winters on the streets would be used to extreme conditions. Laika was trained alongside two other dogs who survived being kept in tiny pressurised cabins for consecutive days and then weeks. Her size and temperament earned her the dubious honour of being chosen to head into orbit.

Sputnik 2 had not been built for retrieval. This was only ever going to be a one-way flight. As Laika was placed into Sputnik 2's tiny capsule, one of the flight technicians kissed her and wished her bon voyage, knowing that she would not survive, despite government PR promises that she would be parachuted safely back to earth. Doctors had embedded monitoring devices into her body to check heart rate, blood pressure, movement and breathing rates.

On 3 November, as Laika orbited the earth, the thermal insulation around the capsule began to come loose. The temperature inside rose to 40C and the sensors showed the dog's pulse rate to be three times higher than its resting level. The exact details of Laika's fate are unclear. At the time, the Soviets said that she had died painlessly after being euthanised with poisoned dog food after several days in orbit. Others said she lived for only minutes. In 2002, one of the scientists involved in the mission told the World Space Congress in Houston that she had been killed by the heat and stress within hours.

Whichever way it happened, it was a murder mission. The dog had been sacrificed in the name of progress. The Soviets ultimately won the race to get a man into space when Yuri Gagarin completed his successful mission in 1961.

Despite protests outside Soviet embassies and the United Nations, four other dogs were sent into space by the USSR to die in Laika's wake. After the collapse of the Soviet regime, one of the men who trained Laika for her fatal flight, Oleg Gazenko, spoke out publicly about his sadness surrounding the little dog's fate:

> Work with animals is a source of suffering to us all . . . The more time passes, the more I'm sorry about it. We shouldn't have done it. We did not learn enough from this mission to justify the death of the dog.

Laika's name and image have lived on in books, cartoons, on stamps and even as a brand of cigarettes. A statue to her stands in the Russian cosmonaut training centre at Star City, near Moscow.

LEARNED PIG

The word 'pig' is hurled as an insult. You're a greedy pig or a dirty pig, never a clever pig or a handsome pig. Pig is a derogatory name for a policeman or the worst of male chauvinists. Selfish drivers are road hogs. An untrustworthy, unfaithful or unkind person is a swine. It's really not fair. Pigs are thought of as smelly, dirty and stupid but that couldn't be further from the truth.

Pigs are clean. Forget the stereotypes. Given enough space, they will always avoid soiling the areas where they sleep or eat. They wallow in mud only to cool themselves down, as they have no sweat glands – which, by the way, brings me to another fallacy: you can't possibly 'sweat like a pig' if a pig doesn't sweat.

Pigs are intelligent. Studies in the 1990s at Emory University, Atlanta, showed that they were able to use a cursor on a computer screen to distinguish between images they had seen before or were seeing for the first time. They learnt to do this just as quickly as chimpanzees.

Pigs are also able to move mirrors to search for hidden food. They can collaborate and communicate through symbols. They have excellent memories and are able to recognise specific individuals. They are not misled by people dressing in identical outfits or changing other attributes of their appearance.

Pigs are emotional. In 2015, researchers at Wageningen University in the Netherlands proved that pigs can share their feelings. They have displayed empathy and sympathy. They care.

One famous example of an outstandingly clever pig was the Georgian wonder, the Learned Pig, who wowed audiences across the country in the 1780s. The pig had been trained by Samuel Bisset, proprietor of a travelling novelty show, and exhibited with great success in Dublin. When Bisset died, the pig passed into the hands of a man called John Nicholson who, according to 'All Things Georgian':

> possessed a peculiar power over animals; he taught a turtle to fetch and carry, a hare to beat a drum with its hind feet; he taught six cocks to perform a country dance; his three cats to play several tunes on the dulcimer with their paws and to imitate Italian opera.

His pig, however, dwarfed all of the menagerie's collective achievements.

Nicholson 'taught' the pig how to tell the time and count the number of people in a room. He could spell out a name and answer direct questions. As one newspaper described it:

> This entertaining and sagacious animal casts accounts by means of Typographical cards, in the same manner as a Printer composes, and by the same method sets down any capital or Surname, reckons the number of People present, tells by evoking on a Gentleman's Watch in company what is the Hour and Minutes; he likewise

> tells any Lady's Thoughts in company, and distinguishes all sorts of colours.

Nicholson took his 'Wonderful Pig' on tour. The crowds flocked to see the spectacle in Leeds, Wakefield, Derby, Nottingham and Northampton. Nicholson netted over 100 guineas a week from his 'grunting philosopher' and the newspapers spilled over with comic superlatives. Samuel Johnson never saw the Learned Pig, but he is said to have read a report of a Nottingham show and commented, 'The pigs are a race unjustly calumniated. Pig has, it seems not been wanting to man, but man to pig.'

All was going swimmingly until April 1785, when the pig and his owner were invited to perform at a private exhibition at Brooks's, one of the most eminent gentlemen's clubs in London. According to a newspaper report,

> A good deal of confusion arose to the master of the pig and the company present, from the improper questions which were put to this grunting philosopher. He counted the company well enough; but when he was asked how many Patriots were present, snorted at every member, and looked around for fresh order . . . 'How many are there present who are six pence clear of encumbrances?' The pig stood still. 'How many honest gentlemen?' The pig would not stir. Here the master was obliged to apologise and in a confounded passion whipped the pig and beat a hasty retreat.

On reflection, I wonder if the Learned Pig read his audience with perfect accuracy. He may have been even cleverer than Nicholson realised. His bacon saved, he trotted off to Europe to play to packed houses across the Continent.

What happened after this is a matter of speculation. Some say that the pig died in 1788. Others that he survived the French Revolution in 1789 to make a triumphant return to his home soil, whereupon he was ready 'to discourse on the Feudal System, the Rights of Kings and the Destruction of the Bastille'.

The Learned Pig proved to be something of a trendsetter. Others to follow in his footsteps included Toby the Sapient Pig who was the talk of London in the early nineteenth century. According to his owner, illusionist Nicholas Hoare, Toby could 'discern a person's thoughts', something 'never heard of before to be exhibited by an animal of the swine race'. He published an autobiography 'written by himself' in 1817.

LIN WANG

Elephants have the largest brain of any land animal, four times the size of a human brain. A number of studies have been conducted into the extent of their cognitive abilities, with results showing that they can demonstrate empathy, work out how to use tools to help them reach food, recognise their own reflections and more.

Research at the University of Sussex has determined that elephants are able to distinguish ethnicity, gender and age simply by listening to someone's voice. Behavioural studies carried out at the University of St Andrews have also proved that elephants have amazing memories and are able to track and recognise up to thirty members of their herd.

Psychologist Richard Byrne has said that their working memory is 'far in advance of anything other animals have been shown to have, adding: 'Imagine taking your family to a crowded department store and the Christmas sales are on. What a job to keep track of where four or five family members are. These elephants are doing it with thirty travelling-mates.'

Elephants, the saying goes, never forget and that really is saying something as, in animal terms, elephants are known for their longevity. Their average lifespan is fifty to seventy years, but one elephant famously lived longer and experienced more than any other. His name was Lin Wang and he became one of Taiwan's greatest folk heroes.

Born in Burma in 1917, Lin Wang was captured by the Japanese Army who used him to haul heavy artillery and supplies through thick jungle, over mountains and across rivers during the bitter Second Sino-Japanese War.

Tensions had been building between Japan and China since the Japanese invaded Manchuria in 1931. The Japanese captured Beijing, Nanjing and Shanghai, while the Chinese fought back with aid from the Soviet Union and the United States. After the attack on Pearl Harbor in 1941, and the USA's declaration of war on Japan, the battle was subsumed into the wider conflict of the Second World War.

When the Japanese attacked British colonies in Burma, the Chinese Expeditionary Force (CEF) was formed under General Sun Li-jen to fight in the campaign and to complete the Ledo Road which ran between Assam and Yunnan, allowing Allied forces to deliver supplies to help the war effort.

In 1943, CEF troops raided a Japanese camp and captured

thirteen elephants, among them Lin Wang. Two years later, when they were recalled from Burma, the elephants went too, on the eighteen-month march to reach Guangdong. Almost half the elephants didn't make it. By the time the survivors arrived at their destination, the war was over. Four of the surviving elephants were sent to zoos across China, while Lin Wang and two others found a temporary home in a park in Guangzhou on the Pearl River, northwest of Hong Kong.

Lin Wang with General Sun Li-jen in Taiwan.

When General Sun Li-jen was sent to Taiwan in 1947, he took the three Guangzhou elephants along with him to his new base. For four years, Lin Wang worked hard, carrying materials for the new railway line. By 1951, Lin Wang was the sole survivor of the original group, and the army decided that he might be better off elsewhere. In 1952 they donated him to Taipei Zoo where he found a mate in Ma Lan, a four-year-old female from Japan.

At thirty-five years old, Lin Wang's war years and public service had earned him the respect of everyone in his new home. He fast became the most popular animal in the entire country and was revered as a cultural icon. As Taiwan moved forward from the austerity of the post-war years and found new prosperity, so respect for their elephant war veteran grew ever stronger. Lin Wang's birthday was celebrated every year with a party at the end of October. Thousands of visitors, including the great and the good of Taipei, would turn up to wish him well.

As the years passed, his great age led people to dub him 'Grandpa Lin'. To celebrate his eightieth birthday, in 1997, a new tropical forest enclosure was constructed in his honour at the zoo. But Ma Lan's death in late 2002 broke Lin Wang's heart and he never quite recovered. He died in February 2003.

The elephant had become such a part of the national identity that the whole country mourned. His 'memorial service' lasted several weeks and 180,000 people came to watch. The president sent a wreath 'to our forever friend, Lin Wang', and the elephant was posthumously made an honorary citizen of Taipei by the mayor Ma Ying-jeou who said, 'Lin Wang was part of the collective memory of four generations of people in Taiwan. He'd seen us growing up, and we'd seen him growing old.'

LONESOME GEORGE

Everyone loves a tortoise, right? Especially a giant tortoise.

Lonesome George was a one-off, a giant tortoise who became a conservation icon. Every island in the Galápagos had its own sub-species of giant tortoise and he was the last known Pinta Island tortoise on the planet. He died at the estimated age of 112 (although Sir David Attenborough thinks he may have only been in his eighties).

Pinta had once been hopping with giant saddleback, long-necked tortoises but goats had ruined the party. Three of them had been shipped to Pinta in 1959 and, in the space of ten years, had grown to a herd of 40,000. The goats gobbled and trampled the vegetation, leaving little on the twenty-three square miles of island for the tortoises to enjoy. They were all thought to have disappeared until Hungarian scientist Josef Vagvolgyi, who was studying snails on the island, spotted George in 1971. In the spring of 1972, Galápagos National Park rangers from the Charles Darwin Research Station found him again and moved him to the tortoise centre on Santa Cruz for his own safety.

A search for other Pinta Island tortoises was launched, both on the island itself and in zoos far and wide, but there were none to be found. The species was pronounced functionally extinct, as George now had no chance of finding another Pinta mate with whom to procreate.

A plan was formed to create a hybrid so that at least some Pinta DNA would survive. A rather overweight George was put on a diet. Two beautiful female *Chelonoidis becki*, from

Isabela Island, were introduced to his corral. Later, two *Chelonoidis hoodensis* from the Ecuador-funded Española breeding programme were also gently sent his way.

George had other ideas. He was happy being lonesome and failed to fertilise any eggs. In fact, all attempts to get George to breed proved futile.

The end of the Pinta Island tortoise came on 24 June 2012, when the now famously celibate George was found dead by Fausto Llerena, the ranger who had looked after him for over four decades. A post-mortem showed that he had died from natural causes. News of his death made headlines around the world.

Lonesome George's body was frozen and sent to New York to be preserved. He was the star attraction in an exhibition at the American Museum of Natural History and, nearly five years later, after a row over whether he should be on display in the Ecuadorian capital Quito or in the Galápagos, he arrived back on Santa Cruz Island.

In the words of the Galápagos Conservancy President Johannah Barry,

> George symbolised the precarious state of biodiversity around the world. He was a catalyst for the extraordinary efforts of the Ecuadorian government and an international network of scientists and conservationists who have undertaken efforts to restore tortoise populations and to improve the status of other endangered and threatened species in the archipelago. He also became a symbol of the tremendous advances that can be made when science, conservation expertise and political will are aligned on a common cause.

Today, visitors continue to flock to see the body of Lonesome George, which lies in state at the Darwin Research Station on Santa Cruz Island, where he had been protected since 1972.

MAGIC

Horses are renowned for their empathy and kindness. Think of the dependable, docile ones who carry children every day at Riding for the Disabled classes and always make sure they feel safe and secure. Miniature ponies can be naughty little things but, in that small package, they can also carry a huge heart.

Strebors Black Magic On Demand (aka Magic) is the perfect example.

With a black body, wide white face and blue eyes, Magic stands out from the crowd even if she doesn't stand more than twenty-seven inches tall. Based in north Florida, Magic has brought comfort and support to those in need across the USA. She was there to help the children and first responders of the shooting at Sandy Hook Elementary School in Newtown, Connecticut, in 2012; she supported survivors of the Moore tornado in Oklahoma in May 2013; and survivors of the horrific attack on the Emanuel AME Church in Charleston, South Carolina, in 2015. The therapeutic use of horses goes back thousands of years. Hippotherapy (*hippos* is Ancient Greek for horse) is mentioned by Hippocrates (460–370 BC), the founding father of medicine and author of the Hippocratic Oath. Horses are sociable herd animals and creating a bond with them can be hugely beneficial. Working with them and caring for them can improve self-esteem, general well-being and self-confidence, as well as concentration, coordination and peace of mind. And horses 'mirror' behaviour, meaning that the calmer and more open you are with them, the more you are likely to get back.

Hippotherapy achieved recognised status in the mid-twentieth century. Riding for the Disabled was formed in 1969 to help children and adults with physical and intellectual challenges. Having seen a lot of their work for myself, I can testify to the confidence gained and the joy shared by those who might never have been around horses before. I have also spent time at a centre in North London called Strength and Learning Through Horses and have seen how effective horses are at working with young people who are struggling with education or feel socially excluded. Equine-assisted therapy (EAT) can help with anxiety, depression, PTSD, eating disorders, brain injuries, musculoskeletal issues and substance abuse.

Generally horses are large, which might be intimidating for many people, although overcoming any fear can be an excellent first step to gaining confidence. Magic, however, has all the qualities of a therapy horse without the size. She is a pocket-sized pick-me-up. She is a member of Gentle Carousel Miniature Therapy Horses, a Florida-based charity founded in 2006 by Jorge and Debbie Garcia-Bengochea. They realised through their own children that miniature ponies have a very special power to connect, and they now ensure that their twenty-six tiny ponies can respond to any emergency.

Magic's journey to Sandy Hook in December 2012 – over 1,000 miles – was longer than most. Despite the chill of a

Connecticut winter, the queue at the Newtown library kept on growing as more than 600 people turned up in search of comfort. For more than two weeks, Magic spent time with the schoolchildren of Sandy Hook, their families and the first responders who had worked at the scene. She would trot up to the children and lay her nuzzling head in their laps, giving them the strength to face the difficult days ahead. It made an enormous difference to them all.

Closer to home, Magic brings solace to many patients in care homes, hospitals and hospices across Florida. She even inspired one woman in an assisted-living facility to find her voice. Kathleen Loper had not uttered a word for three years but when she saw the blue-eyed miniature mare she said, 'Isn't she beautiful?' From that day onwards she could communicate again.

MICK THE MILLER

A new sport emerged in the 1920s that gave the British a chance to 'go to the dogs' in a good way. Against a backdrop of economic depression and soaring unemployment, greyhound racing provided much needed cheer.

Early in its rise to prominence, the sport benefited from a superstar who captured the attention and support of the masses: Mick the Miller was the first great in the world of English greyhound racing. In the fame stakes he was right up there and could draw tens of thousands to watch him race.

The pale brindle Greyhound was bred by a gambling priest called Father Martin Brophy in Killeigh, Co. Offaly, Ireland, in June 1926. He was named after the odd-job man who helped out at the vicarage. Although Mick was the smallest and weakest of his litter, he came from good stock, a direct descendant of Master McGrath, winner of the prestigious Waterloo Cup. This trophy was the highest honour in the world of coursing, where greyhounds chased a real hare across the countryside.

Michael Greene, who worked for Father Brophy, picked Mick and his brother Macoma out of the litter and asked if he could rear them. He gave them constant attention, often feeding them from a bottle and even letting them share his bed. He also gave them a regular routine, walking them for miles to build up their muscle and paving the way for a professional racing career.

The timing was serendipitous. Stadium racing, with dogs going round an oval track chasing a mechanical version of a hare, had just started in Ireland and the USA. It would shortly reach the UK.

When Mick displayed early potential, Father Brophy negotiated for him to join the stable of the biblically named Moses Rebenschied. However, a series of dramatic events changed the course of Mick the Miller's career. A tornado blew the roof off Rebenschied's kennel in St Louis, Missouri, killing twenty-seven of his greyhounds. The tornado also overturned a van driven by his son, leading to the loss of another four. Moses declared, 'The hand of God is warning me against greyhounds.' The deal was called off and Father Brophy returned the cheque.

That twist of fate meant that Mick the Miller stayed closer to home. He was sent to a trainer, Mick Horan, based at Shelbourne Park racetrack, near Dublin. In his debut season, he won four of his five races and set tongues wagging when he equalled his brother Macoma's world record of 28.80 seconds for 500 metres.

Then, in May 1928, shortly before his second birthday, Mick was diagnosed with distemper. It could have proved fatal but luckily the manager of Shelbourne Park was also a qualified vet and saved his life. Whether he would recover well enough to fulfil his racing potential was uncertain.

When Mick the Miller was well enough to race again, he soon showed he had lost none of his pace. Four easy wins in Ireland convinced Father Brophy that he was the dog to venture across the Irish Sea in search of greater riches. The Greyhound Derby at White City was the ultimate prize.

In a solo trial Mick the Miller broke the White City track record. Suddenly, from being a 25–1 outsider, the dog no one had heard of was the 4–7 favourite.

He won the first round by eight lengths, also setting a new world record over 525 yards, becoming the first dog to break the magic thirty-second barrier by clocking an impressive 29.8 seconds.

Rather unromantically, Father Brophy cashed in and sold him for 800 guineas to a bookmaker from Wimbledon called Albert Williams. It was a huge sum of money, enough to buy a house in most areas of London. Part of the deal was that Father Brophy would get any prize money Mick earned on the night and that he would keep the trophy if he won.

Forty thousand spectators watched Mick and his three opponents line up for the final at 8.45pm. There was a collision on the first bend and Mick the Miller was beaten into second place, but the race was voided. The rerun was half an hour later and Mick won by three lengths. When news of his victory reached his home town, the people of Killeigh held an impromptu party in his honour.

By the end of 1929 he had won twenty-six of his thirty-two races and found a new home after being sold to Arundel H. Kempton for a jaw-dropping £2,000 as a gift for his wife Phyllis.

Mick the Miller receives a massage at his kennels.

Mick's successes continued to pile up. In 1930, he won twenty out of twenty-three races, including the English Greyhound Derby (for the second year running), the Spring Cup at Wembley, the Cesarewitch at West Ham and the Welsh Greyhound Derby. He broke world records on four occasions. Headlines described him as a 'wonder dog' and 'invincible'.

Sadly, a torn shoulder muscle during a race marked the beginning of the end of his exceptional career. For the first time he lost three races in a row, though he still managed to qualify for the final of the Derby. Seventy thousand spectators looked on with bated breath as Mick and the 'Black Express', Ryland R., stepped up as joint favourites to take on four other dogs on 27 June 1931. With six runners in the final, it was bound to be a rough race.

As they rounded the first bend, Ryland R. was well ahead, with Mick trailing in last place. Until the final turn, Ryland R.'s victory seemed assured until he snapped at a rival. Trying to bite the opposition is a sin punishable by disqualification and the klaxon sounded to signify a void race but, at the same time, the crowd were roaring because they could see Mick the Miller making ground on the inside rail. They roared him home as Mick finished with a heroic surge and put his nose in front on the line.

Ryland R. was disqualified for 'nosing and impeding', but the crowd made their disapproval clear as the announcer repeated that it was a void race. Mick's owner, Phyllis Kempton, broke down in tears, crying, 'Mick has won! My darling Mick has won!'

With Ryland R. out and Kempton refusing to allow Mick to race on the basis that he had already run and won, the stewards of the Greyhound Racing Association had a situation on their hands. They finally talked Mrs Kempton round, although Mick, without the recovery rate of his youth, was at a huge disadvantage. He couldn't write the fairy-tale ending twice and trailed in fourth.

The trophy was awarded to the winning owner amid a chorus of boos from the crowd.

The record books might not show Mick the Miller as the winner of his third and final Derby, but in his wrongful defeat, his star had never shone brighter.

'Greyhound racing is still in its infancy', declared the *Greyhound Mirror and Gazette*, 'but already it has produced a popular favourite as idolised as any horse, cinema star, footballer or boxer in history.'

By now a true global superstar, Mick continued to race throughout his final season, finishing his track career with an outstanding victory at the St Leger Stakes in front of 40,000 spectators in a race later described as the 'greatest ever to be held at Wembley's Empire Stadium'.

After his retirement from racing in December 1931 he opened shops, attended high-profile events and rubbed shoulders with royalty. He became the most expensive dog advertised at stud (at 50 guineas a go) and amassed a fortune of £20,000 in stud fees, appearance fees and prize money.

A life-size statue of Killeigh's most famous resident stands proudly in the village and, even now, Mick is still known as the world's most famous greyhound, the only dog ever to have won the treble of the Derby, Cesarewitch and St Leger.

The distinctive, sleek physique of a greyhound is built for speed. Greyhounds are the fastest breed around, reaching up to 45 miles an hour, faster than a racehorse. Their acceleration is extraordinary (0–30mph in just three seconds) and their widely spread eyes give them 270-degree vision (humans have only 180 degrees), meaning they can see some of the back of their head. Although they stand up to 30 inches tall, they are incredibly lean, weighing only about 60–70 pounds.

Greyhounds have been around for over 3,000 years. In Ancient Greece, hunters would use them in coursing competitions to chase down prey. In 1014, King Canute ruled that only noblemen could own them. They were deemed to be more valuable than serfs, and killing a greyhound was a crime punishable by death. Henry VIII and Elizabeth I were also fans (of both dog and sport).

MILTON

'Milton was the special one – he was the horse of a lifetime', says John Whitaker, one of the greatest show jumpers who has ever lived.

I used to dream about Milton. I had posters of him on my wall, soaring over massive parallel bars. He was like Pegasus, able to fly. He had a presence and an elegance unlike other horses. His galloping action was so fluid and natural that he was like a dancer floating across the ground and when he jumped, he sprang upwards and forwards with his front legs outstretched in a way I had never seen a horse move before. He was also strikingly handsome, a pure white grey with a long flowing mane. He was the stuff of fairy tales, and yet he was real.

Milton was born in April 1977 and bought as a foal by the brilliant show jumper Caroline Bradley, who marked him out as a future Olympic horse. She had ridden Milton's sire Marius to win the Queen Elizabeth II Cup at the Royal International Horse Show and she hoped his son would have a similar talent. Tragically, Caroline died of a heart attack at the age of only thirty-seven. She did not get to ride Milton in his prime but her family maintained ownership of the horse.

The Bradleys asked John Whitaker to take over as Milton's rider in 1985, when the horse was eight years old. Like their daughter, Whitaker had beautifully soft hands and a quiet way in the saddle that they thought would suit their talented but wilful horse. You see, Milton was not straightforward. His long stride meant that he didn't like to 'shorten up' (horsey speak for putting in quicker, little paces) into a jump. He needed to be allowed to flow to be at his best.

Whitaker had won team silver at the Los Angeles Olympics the previous year, with the brilliant Ryan's Son, with whom he'd been in partnership for fourteen years. His first impression of Milton was that his front legs were 'dangly', but he still had star quality. 'When he takes off you feel that you'll never come down', he said when he first jumped him, and explained, 'His back is too powerful for his front. You have to give him time and room.'

He was patient in getting to know the beautiful grey horse, shortening his stirrups so that he could sit out of the saddle and be lighter on his back. He realised that Milton did not like to be controlled by a strong leg or an iron hand. The horse needed to feel he was in charge. The two formed an unbeatable partnership and became part of the hugely successful British show-jumping team.

In 1986 they won the Du Maurier Limited International at Spruce Meadows in Canada, then the competition with the world's highest prize money. A year later at the FEI European Championships in St Gallen, they took team gold and individual silver. They repeated this in Rotterdam in 1989.

Whitaker and Milton were first pick for the British team to go to the Seoul Olympics in 1988, but controversially they stayed at home. The Bradleys didn't want Milton to travel so far to compete in a hot and humid country. Despite the immense pressure put on them (which included not allowing the horse to take part in the Nations Cup that year), they stood firm. They may have been right and, in protecting their horse, ensured that Milton stayed at the top longer than most.

Milton was strong-minded too. If he didn't want to move then he wouldn't. If there was something he didn't fancy doing, he would blow raspberries in protest. He also had a predilection for removing his rugs and shredding them to bits. Fortunately none of these peccadilloes stood in the way of his genius.

Whitaker and Milton became the top jumping pair in the world as the medals continued to roll in. Top of the pile was victory at the Longines FEI Jumping World Cup final in 1990 and 1991, which marked him out as the best horse in the world. Individual silver and team bronze medals came at the first FEI World Equestrian Games in Stockholm, 1990. At the Horse of the Year Show there was a competition called the Masters where the fences got higher and higher every round. Milton won it three years in a row without having a single fence down.

In the FEI Jumping Nations Cup, he jumped an extraordinary thirty-five clear rounds and twelve double clears in a seven-year period. Milton could jump anything. There was no course too big or too difficult for him. The crowd loved him and would roar their approval as he leapt into the air in celebration of another clear round. 'Milton loved all the attention – the big crowds, the big atmosphere. The more atmosphere and noise, the better Milton was', Whitaker says.

Milton confirmed the prediction of Caroline Bradley by going to the Barcelona Olympics in 1992, but things didn't go to plan. He jumped clear in the first round of the individual competition, but in the second he stumbled in the deep sand in the middle of a double of parallel bars and Whitaker had to stop him from jumping the second part. They picked up three faults for the refusal and finished with fifteen faults which gave them no hope of a medal. 'I would say he fractionally hesitated, possibly he was trying to be too careful', reflects John, and continues: 'The ground didn't suit him a lot. Afterwards, he gave up a little bit and I gave up a little bit, we lost concentration and we had another three rails down.'

Although Olympic success eluded him, Milton's exceptional talent and consistency made him show jumping's first millionaire and the first non-racehorse to win over £1 million in prize money. He was also a sponsor's magnet and carried various pre-names including Henderson, Everest and Next.

Milton finished his career in 1994 with a grand farewell at Olympia. He made a few guest appearances at shows, some with the other dashing grey of the equine world, Desert Orchid, but mainly he lived out a happy and quiet retirement at Whitaker's farm in Yorkshire. He died at the age of twenty-two and was remembered as a horse who had the talent and the profile to have made show jumping one of the most popular and marketable sports in the UK. John Whitaker is still competing at the age of sixty-five.

MINNIE

I mentioned in the Introduction that my first pony was a Shetland called Valkyrie. She was small, round, furry and full of character. She had spent her early life in the grandest surroundings at Windsor Castle, where she had taught Prince Andrew and Prince Edward to ride. She arrived at Kingsclere in 1971 not long after I was born, as a gift from HM The Queen. After she had tried to teach me to ride and schooled me in how to behave properly, she lived on with us until she was in her thirties. Whenever the Queen came to see her racehorses, she also loved to see Valkyrie. Hence a long line of beautifully toned and groomed thoroughbred racehorses would be joined by a fat little Shetland pony.

'Ah, Valkyrie!' the Queen would beam with delight. 'She looks so well.'

I am pretty certain that Valkyrie recognised the Queen too and bowed her head deliberately. There is no doubt that Her Majesty never forgot her.

This is the tale of another small but memorable pony. It takes us back to the Second World War, when the Chindits, special operations units of the British and Indian armies, were fighting the Japanese Army in North Burma. There was no real hope at this stage of driving the invaders out of the country, but the soldiers formed a secret 'phantom army' whose mission was to launch repeated surprise attacks on the Japanese camps and bases before disappearing back into the jungle.

It was a tough campaign. The terrain was difficult and the troops, weakened by hunger and disease, faced lengthy marches in their bid to head off the enemy. Many were killed and wounded. Some suffered so badly with dysentery that they had to cut holes in the seat of their trousers to avoid having to make constant stops on these arduous treks. Morale was low.

At the White City compound at Mawlu, men took great comfort in the animals in their midst, especially one. During a particularly brutal attack by the Japanese forces in which many were killed, one of their mares gave birth to a foal. The troops named her Minnie and were delighted to have a new distraction from the death and destruction that surrounded them. Whenever there was a lull in the fighting, more and more of the men came to see the spindly-legged little foal for themselves. She rapidly became their unofficial mascot.

The bombardments continued, and in another deadly raid a panic-stricken mule kicked Minnie in the eye. The men did what they could to save the creature; and such was the concern that 77th Brigade Commander Mike Calvert (affectionately known as 'Mad Mike') ordered that regular reports of the pony's progress be circulated to all forward positions. Fortunately, Minnie made a good recovery and took to visiting the soldiers manning the various mortar positions to scrounge sugar lumps and tea, which she drank from a pint pot.

When the troops eventually evacuated White City, Minnie was deemed too young to manage the long march through hostile jungle, so the brigadier, in gratitude for what she had done for his men's morale, arranged to have her flown to India, despite the obvious dangers. The use of the plane, which would have to fly deep into enemy territory without being spotted, endangering both men and machines, broke just about every law of military discipline. And yet Minnie was worth it to the Chindits.

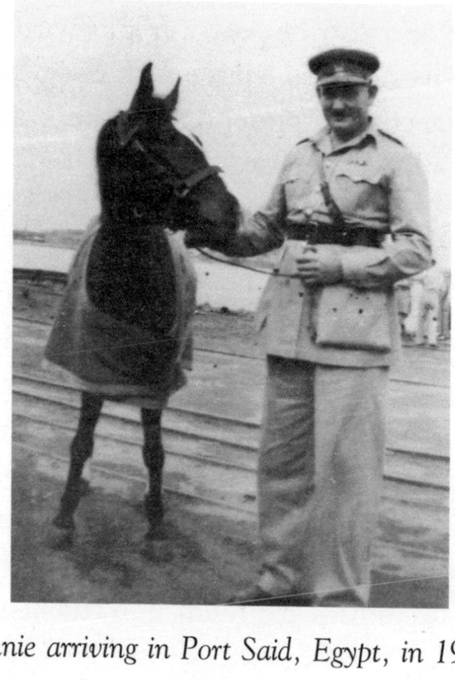

Minnie arriving in Port Said, Egypt, in 1950.

She remained in Dehra Dun until the battalion returned home in 1944 and continued to make her mark on those around her, not least by eating the tablecloths in the sergeants' mess. After the war, Minnie was allowed to remain with the Lancashire Fusiliers, and became their official mascot. She travelled back to England with the troops on the *Georgic* in October 1947, enjoying the sea air and continuing to eat anything she could snaffle from her fans, including Spam and condensed milk.

Back in England, and stationed in Shropshire, Minnie took part in ceremonial parades dressed for the part in a special bridle and saddlecloth. When the battalion moved to Warminster, Minnie went too. She took part in Trooping the Colour in 1948 and travelled to Egypt on a tour of duty before succumbing to pneumonia in November 1951.

Minnie was (mostly) buried in the camp at Moascar, although two of her hooves were made into an inkwell and a paperweight and can be seen on show with her tail and blanket at the Regimental Museum in Bury, Lancashire. Her other two hooves were presented to the boroughs of Bury and Rochdale in her memory.

MIRACLE MIKE

At times when we are busy or stressed, we've probably all used the phrase 'running around like a headless chicken'. For one cockerel in the USA, running around like that is exactly what he did for eighteen months. This is the story of Miracle Mike.

It was September 1945 on a small farm in Fruita, Colorado, and it was time to slaughter poultry. Lloyd Olsen and his wife Clara could work through forty or fifty birds in a single session. Lloyd would do the decapitating and Clara would clear up the mess.

It is often the case that a chicken will be able to walk for around fifteen minutes after it has had its head chopped off because there is still residual oxygen in the spinal cord circuits and the neurons can activate automatically, causing the legs to move. The chicken can run around with no aim or understanding, eventually dropping dead when the oxygen is exhausted.

In this case, the chicken who walked around after his head had been chopped off did not drop dead within the hour, nor within the day, the week or the month. Instead, he gained celebrity status in post-war America.

Miracle Mike at home in the barnyard with the other roosters.

Olsen left the decapitated cockerel in a box, not expecting that it could survive the night. However, the next morning the rooster seemed as active as ever. Olsen, sensing an opportunity, took him to the market along with his (definitely dead) coop mates. He bet people that he had a live headless chicken in his wagon. He won every wager.

Word soon spread and the news appeared on the front pages of local newspapers. Then the story reached a wider circle and caught the eye of a man called Hope Wade, a sideshow promoter from Salt Lake City, Utah – think of him as the Don King of his day. He could create a crowd around a paper bag so give him a real, live, headless chicken and there was serious money to be made. Understanding the draw of a good name, Wade called the cockerel 'Miracle Mike, the Headless Chicken'. *Life* magazine picked up the story and a legend was born.

Before heading out on tour, the rooster was subjected to tests by scientists at the University of Utah, who then tried removing the heads of a number of other chickens to see whether they too would survive (they didn't).

So it was that the Olsens, Miracle Mike and Hope Wade headed west to California, where the public rushed to pay the twenty-five-cent admission fee to see the spectacle of a living beast without a head. Clara kept all the details of the tour and the newspaper coverage in a scrapbook.

'How did the chicken survive if it couldn't eat or drink?' I hear you say.

Well, that is a very good question but the Olsens were farming folk and they knew how to keep an animal alive if needs be. They dropped liquid food and water directly into Miracle Mike's oesophagus and removed any mucus with a syringe. As long as they had both instruments to hand, all was well. Mike carried on trying to preen, peck and crow, though his attempts at the latter sounded rather more like a gurgle.

Also key was that he could still think (as much as a chicken does think), or at least his brain could still send messages to the key functioning elements of his body. The angle of the axe that had cut off his beak, eyes, ear and face had left intact most of his brain, 80 per cent of which sits very low and to the back of a chicken's head. How he didn't bleed to death seems to have been down to the miracle and the placement of a well-situated and well-timed blood clot.

The money kept on coming. They may not have made a fortune, but the headless cockerel provided enough for the Olsens to replace their horse and mule with two tractors, a hay baler and a 1946 Chevrolet pick-up truck. In addition to the financial gain, the Olsens were having the adventure of a lifetime, seeing parts of the States they would never have had the opportunity to visit.

Then it all went wrong. One night in the spring of 1947, while on tour in Phoenix, Arizona, Lloyd and Clara were woken in their motel by the noise of choking. Miracle Mike needed his throat cleared. Lloyd searched for the syringe, only to find he had left it at the showground.

Miracle Mike choked to death.

The Olsens' great-grandson Troy Waters later told the BBC that Lloyd could barely bring himself to own up to what had happened:

> For years he would claim he had sold [the cockerel] to a guy in the sideshow circuit. It wasn't until, well, a few years before he died that he finally admitted to me one night that it died on him. I think he didn't ever want

to admit he screwed up and let the proverbial goose that lays golden eggs die on him.

The legend of Miracle Mike lives on, at least in Colorado. On the third weekend of May, Fruita hosts a 'Mike the Headless Chicken Day', which includes activities such as 'pin the head on the chicken' and a five-kilometre 'run like a headless chicken'. He also inspired the 2008 song, 'Headless Mike', by the Radioactive Chicken Heads.

MOKO

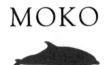

This book is not only about animals whose heroics save us humans, make us better people or sacrifice themselves for our benefit. I'd also like to pay tribute to those who help their fellow animals in times of need. Moko the dolphin is one such saviour, swimming into action just in time to save the lives of two fellow sea dwellers. Think *Baywatch* without the bathing suits and you'll get the picture.

Moko lived in the waters off Mahia beach on New Zealand's North Island for over two years. During the summer he loved to swim with visitors, often stealing their surfboards or kayaks. In March 2008, two pygmy sperm whales got trapped between a sandbar and a beach on the Mahia peninsula. For ninety minutes, locals did everything they could to refloat the whales but nothing seemed to be helping. Rescuers started to think that putting them down might be the only way to save the pair from a distressing and painful death.

Suddenly, the Bottlenose dolphin appeared. Succeeding

where humans had failed, he somehow communicated with mother and calf and led them through a narrow channel and out to the safety of the open sea. A local conservation officer told the BBC:

> I don't speak whale and I don't speak dolphin, but there was obviously something that went on because the two whales changed their attitude from being quite distressed to following the dolphin quite willingly and directly along the beach and straight out to sea.

Moko, named after Mokotahi, a headland on the peninsula, was already well known in the area and now he became an even bigger attraction. He thrived on company, becoming bored in the winter months when fewer people were around. One winter swimmer found this out to her cost when she braved the waters and started to play with the dolphin. When she tired, he wanted to carry on and he blocked her from returning to shore. After she had been rescued, she admitted that the dolphin had meant no harm, and it was probably not wise of her to go out late and on her own.

In September 2009, Moko moved up the coast to Gisborne where he again won hundreds of fans, playing with swimmers in the sea and in the river system where he seemed really to enjoy being petted as well as stealing balls and boogie boards.

His travels up the northeast coast were not without incident. He sustained an injury on his upper-right jaw from a fish hook and various knocks from boats along the way. After he followed a fishing boat north to Tuaranga, there was concern about his welfare and his changing behaviour. Scientists had identified that almost half of the 'lone' dolphins who sought human interaction would die before their time.

Moko's body was found on a beach at Matakana Island a few weeks later. The cause of death has never been confirmed but it is believed he may have drowned in set nets. There was national mourning as the news broke. Hundreds of people turned out for his funeral procession and memorial service before he was buried in the Maori tradition on the beach where he was found.

MOLLY

To the list of great crime-solving duos – Starsky and Hutch, Cagney and Lacey, Mulder and Scully, Holmes and Watson – add Colin and Molly, the UK's first pet detective agency.

Colin Butcher is a former police officer, a detective inspector who also worked in the drugs enforcement unit, and Molly is a Cocker Spaniel, a former rescue dog who now spends her life recovering lost cats. I met them both at Crufts when they came into the TV studio and we got a huge reaction to the interview, partly because of Molly's success rate and partly because it contradicts the traditional idea of dogs chasing cats away. And with more than 100,000 cats reported missing in the UK over the last decade, Molly is in huge demand.

Colin has first-hand experience of the anxiety caused when a cat goes missing. When he was a child, he remembers searching fruitlessly for the family cat, Mitzi. Gemini, the dog, kept scratching at the floor in a certain spot, and they eventually realised that, while Colin's father had been repairing pipework under the floorboards, Mitzi must have sneaked down there and become trapped. Mitzi was recovered unharmed. Gemini's crucial role in the rescue sowed a seed in Colin's mind and the concept of a cat-detection dog was born.

When he left the police force, Colin was finally able to bring his long-held plans to fruition and in 2005 he set up UKPD – the United Kingdom Pet Detectives agency, initially focussed on recovering stolen horses and dogs. Colin soon noticed that 50 per cent of the enquiries they received were about cats, and that was when he realised that he needed a new, very special partner.

He wanted a Spaniel but it was essential to find one who didn't mind cats. The dog also needed to be intelligent and focussed. He was keen to take a rescue dog, one that had had a difficult start and could now have a new life with a purpose, but finding the right one was not easy.

He assessed around a dozen before he came across eighteen-month-old Molly through an advert on Gumtree with the message: 'Needs a good home. Owner cannot cope'. The tricolour working Cocker Spaniel with hazel eyes and keen, floppy ears, had already had three owners and her unruly behaviour didn't necessarily mark her out as the perfect companion. Yet Colin knew instantly that she was the one: 'I've been around hundreds of dogs and have never seen one that has her focus. She's so good at what she does and will just keep going and going and going.' The fact that Molly is so good with people is also hugely important and helps to reduce the anxiety that owners experience when their beloved pets have disappeared.

There followed months of intensive training at the Medical Detection Dogs academy in Milton Keynes, where Molly learnt to isolate scents and understand signals and commands.

Then came numerous field trials, including 'cat testing' (to make sure she didn't just chase them). Finally, she was ready to begin work alongside her owner.

Molly uses scent matching to track down the missing cats. A sample of their hair gives her enough to follow their trail, ignoring any other cats that might be around. Like any detective worth his or her salt, Molly is fully kitted out for the job. She wears a fluorescent harness and has her very own abseiling kit, used to lower her over walls when needed. When she finds a missing cat she signals to Colin by lying down; this also helps not to scare the cat. Her successes are rewarded with treats, including her favourite – black pudding.

The combination of Molly's nose and Colin's investigative skills has helped to find lost cats all over the UK, including one that was presumed drowned in the Thames after falling off a houseboat. The devastated owners wanted to recover their pet's body, but Colin suspected that the cat had actually swum ashore and, three days later, Molly found him hiding under a caravan.

Molly's record is quite extraordinary but one of her searches almost cost her her life. She was following a trail that took her on to woodland, where an adder, which wasn't able to get out of her way, bit her twice, injecting venom into her bloodstream. She was paralysed almost immediately. Colin rushed her to a vet, but he didn't have the antidote so all they could do was watch and wait for forty-eight hours.

Molly recovered, although Colin was concerned that getting back to full fitness was taking her longer than it should and that she was still walking with a pronounced limp. Sometimes she is too clever for her own good and Molly was rumbled when Colin's girlfriend spotted her walking normally. Suspicions were aroused that maybe she was putting on the limp whenever Colin was around, so he set up a camera to check. Molly was indeed having him on, so her sabbatical was over.

Colin and Molly receive at least fifteen calls a week from distraught cat owners needing help. Colin says, 'Without [Molly] none of this would have happened. She tests me, always surprises me and never lets me down. She is one in a million.'

MR MAGOO

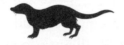

Not many animals can claim to have been saved from execution by John F. Kennedy, but then Mr Magoo the mongoose was unique.

On the northwestern coastline of Lake Superior is the city of Duluth, the westernmost port for transatlantic cargo ships. A lot of cargo comes into Duluth: coal, iron ore, grain, clothing, white goods and, in November 1962, a mongoose from India. The merchant seamen had enjoyed his company on the long journey and had sat drinking tea with him, but they decided he deserved a life on dry land so presented him as a gift to the city's Lake Superior Zoo.

Lloyd Hackl, the director of the zoo, was delighted and named his new mongoose Mr Magoo. It wasn't going to be that straightforward though, was it? This was the USA in the 1960s and people were all twitchy about foreign invaders with their fancy ideas.

The federal agents declared the mongoose an invasive species. They didn't just demand Mr Magoo's deportation, they sentenced him to death. The headline in the *Duluth*

Herald read: 'Mongoose Seized as Undesirable', and continued, 'The Duluth zoo's tea-drinking mongoose has been nabbed by the US government as an undesirable alone.'

The citizens of Duluth were not taking the death sentence lying down. It was pointed out that, as the only mongoose in the country, Mr Magoo was never going to be able to procreate, so the country was unlikely to be overrun by the species.

They demanded he be allowed to live out his days in peace. Petitions were signed and sent to the great and the good, among them the US Secretary of the Interior Stewart Udall, US Senator Hubert Humphrey and Duluth Mayor George Johnson. A campaign, brilliantly nicknamed 'No Noose for the Mongoose', was backed by more than 10,000 citizens. It was suggested that the zoo director, holder of the only keys to the mongoose's cage, should take him into hiding.

Whether or not JFK directly intervened we do not know, but at the eleventh hour, Mr Magoo was pardoned. A statement from Udall read, 'Acting on the authority that permits importation of proscribed mammals – including mongooses – for zoological, education, medical and scientific purposes, I recommend that Mr Magoo be granted non-political asylum in the United States.' He added that this was a very specific case and it did not give free rein to other mongooses entering the country and it was dependent upon Mr Magoo maintaining his 'bachelor existence'.

'MAGOO TO STAY. US Asylum Granted' was the headline in the *News Tribune*. At the thought of his mongoose remaining at the zoo, Hackl told the press, 'I feel wonderful.' As for President Kennedy, he declared: 'Let the story of the saving of Magoo stand as a classic example of government by the people.'

Magoo lived happily in his new home, eating an egg a day, still enjoying a decent cuppa and exercising happily in the zoo office, where he charmed workers with his friendly nature. He was hugely popular with visitors, especially children, and received large numbers of letters and Christmas cards.

When Mr Magoo died peacefully in January 1968, his obituary in the *Duluth Herald* read: 'OUR MR MAGOO OF ZOO IS DEAD'. The new zoo director, Basil Norton, vowed never to replace him: 'Another mongoose could never take his place in the hearts and affections of Duluth people', he said.

Mr Magoo was stuffed and proudly displayed at the Lake Superior Zoo, where visitors can still see him from 10am to 4pm, Thursday to Monday.

NING NONG

It is impossible to forget the events of Boxing Day 2004, when the Sumatra-Andaman earthquake triggered a tsunami in the Indian Ocean. The force was so great – its energy the equivalent of 23,000 atomic bombs – that it caused a shift in the Earth's mass which changed the planet's rotation. Waves reached heights of more than 100 feet and travelled as fast as a jet plane, devastating shorelines of Thailand, Sri Lanka and Malaysia. Almost 230,000 people lost their lives.

Amber Mason from Milton Keynes, aged eight, was on a once-in-a-lifetime holiday in Phuket, Thailand, with her mother and stepfather. Every morning she would rush down to see the elephants outside their hotel. She loved being allowed to ride them along the beach and into the sea; she soon had a favourite, four-year-old Ning Nong. The feeling was mutual. 'He would always grab my hand and pick me out of all the rest', Amber says. She fed him bananas and he would nuzzle her with his trunk. The highlight of her holiday was being able to ride on his back every day.

Boxing Day dawned much like any other day. There was a small earthquake early on, but no one dreamed that it was the precursor of what was to follow. After breakfast, Amber rode Ning Nong along the beach as usual, but there was something different about her favourite elephant that morning. As the tide went out and people ran across the sand to collect the beached fish, including the mahout who was responsible for looking after the elephants, Ning Nong seemed anxious. He was constantly turning away from the sea, rather than towards it as he normally would. He kept pulling away from the mahout, trying to get off the beach.

The elephant sensed something was about to happen. His instinct saved Amber's life. The elephant and his mahout started running away from the sea. Before they knew it, the first waves caused by the tsunami rushed up the beach. As the water rose rapidly, Amber was terrified. She clung to Ning Nong's back as he made his way to higher ground and took her to a stone wall high enough to allow her to climb off his back and stay safe.

The first Amber's mother knew of the unfolding disaster was the sound of screaming coming from the beach. She knew her daughter was with Ning Nong, but the elephant was nowhere to be seen. When someone said they thought he had been killed by the raging waters, she began to panic.

Suddenly she spotted him in the distance, wedged by the wall, while Amber made her way to safety. Grabbing her daughter she made it back to the hotel just in time – minutes later a wave destroyed some of the rooms on the floor beneath them.

Amber knows that without Ning Nong she wouldn't be here. 'He saved my life', she told the *Daily Mail*. 'He knew the signs that something bad was going to happen and he carried me to safety. I will always be grateful.'

Amber's mother has never forgotten how close she came to losing her daughter and sends money to help fund the elephants of Phuket every year.

PADDY THE WANDERER

Dash was an Airedale Terrier given to the Glasgow family in Wellington, New Zealand, in the 1920s. John Glasgow spent a lot of time at sea so Dash was a great companion for his wife and particularly for his little girl, Elsie. The three of them would go down to the wharf to meet John when he was due to sail back in.

When Elsie caught pneumonia and died before her fourth birthday, her family were devastated and Dash took to wandering along the wharf, seemingly searching for Elsie. He never went home again.

Roaming the wharves, the dog soon became familiar to Wellington's seamen and waterside workers. They renamed him Paddy, and together with local taxi drivers, fed him and took turns to pay for his dog licence every year. Paddy was formally adopted by the Wellington Harbour Board and made assistant night watchman, responsible for guarding against 'pirates, smugglers and rodents'.

Paddy the Wanderer, assistant night watchman, at Queens Wharf, Wellington, in 1935.

Paddy then began a wider reconnaissance of the city, riding the tram network. As the Great Depression took hold, he left the New Zealand capital behind him altogether. He stowed away on a ship bound for Australia, visiting various New Zealand ports and eventually returning to Wellington on another boat. His intrepid tales became the talk of the town and Paddy the Wanderer gained a reputation for derring-do. When he visited Auckland it is said the folks working on the water there tried to kidnap him, but fearing retaliation from their Wellington rivals let him travel safely back.

When Paddy became ill in July 1939, the taxi drivers paid for him to go into kennels to recuperate. The dog had other ideas. When one of the drivers went to visit him, he leapt into the back of his cab and refused to leave until he had been driven back to the docks. There his benefactors made a bed for him in one of the sheds, but he died not long afterwards on 17 July.

The *Evening Post* reported that Paddy's coffin, bearing the inscription 'Paddy the Wanderer – at rest', was driven by a procession of twelve taxis and brought downtown Wellington to a standstill as people came out to pay their respects.

In 1945 a collection was launched to fund a memorial near the Queens Wharf gates. The statue, made from bronze and granite taken from London's first Waterloo Bridge, also has a drinking fountain with bowls below for dogs.

PAUL

The octopus is a marvel of a beast. With nine brains, three hearts, blue blood and the ability to camouflage itself, it is among the most cunning and impressive of sea dwellers. However, in 2008, an octopus knocked the ball out of the park by proving himself capable of seeing the future.

Paul was born in Weymouth in 2006 and spent most of his life at the Sea Life aquarium in Oberhausen, Germany. During UEFA's 2008 European Football Championship, he correctly predicted the results of a number of matches and became the guru we turned to for result forecasting.

Two years later he foretold the outcome of *all* of Germany's World Cup matches with pinpoint accuracy, as well as backing Spain to beat the Netherlands in the final – which they did, 1–0 after extra time.

His methodology was to have two boxes of seafood placed

before him, each one bearing the flag of a participating team. The one he ate from first would be the winner.

Inevitably, there were doubters and disparagers. The President of Iran, Mahmoud Ahmadinejad, accused the soccer sensation of being a symbol of Western decadence and decay. I mean, honestly.

The fact remained that Paul succeeded where many pundits failed and became the first – and probably only – octopus oracle with a success rate of over 87 per cent.

As for the science behind Paul's predictions, marine biologists suggested that octopuses are attracted to strong patterns of horizontal stripes. While they are colour blind, they can distinguish brightness, so is it coincidence that the countries picked out by Paul, such as Germany and Spain, have flags that feature bold and vibrant bands?

Others suggested that habit might have led to the octopus picking out Germany's flag time after time. That didn't stop fans believing that the cephalopod truly was clairvoyant.

Paul's predictions became an integral part of the 2010 competition, each one eagerly awaited and broadcast live by a German news channel. When Paul predicted that Germany would beat Argentina in the knockout stages (they did, 4–0), a well-known Argentine chef posted a recipe for octopus on Facebook. However, when Paul forecast Spain's victory over the German team in the semi-final, it was disgruntled home fans who sent death threats and threatened to turn him into sushi.

After the final, when Spain lifted the trophy, their Prime Minister José Luis Rodríguez Zapatero called for bodyguards for Paul and offered him official state protection.

Paul's fame encouraged other German keepers to try to get their own charges in on the act. But none could match up to his prowess. At Chemnitz Zoo, Leon the porcupine and

Petty the pygmy hippopotamus both failed to predict enough correct results. Anton the tamarin ate a raisin representing Ghana, who were then beaten on penalties by Uruguay in the quarter-final. Whether in eating Ghana he was predicting they would be gobbled up or he was selecting them to win . . . well, who knows?

An octopus in Japan did correctly predict all Japan's group stage results at the 2018 World Cup in Russia, and ended up on the slab at a local fish market shortly afterwards.

Despite receiving offers from around the world (one of which was a transfer fee of 30,000 euros), Paul opted for a quiet retirement. Staff at the aquarium presented him with a replica of the World Cup trophy and adorned it with three mussels for him to snack on at his leisure.

Paul died prematurely and unexpectedly in October 2010, prompting a number of conspiracy theories. He was remembered as an octopus who had 'enthused people across every continent'. Twice during the 2014 World Cup, he was featured as a Google doodle, once atop billowy clouds in heaven and wearing a halo and, on the day of the final, cheering on the teams from his celestial resting place.

PICKLES

Paul the Octopus is not the only animal to have found fame through association with the FIFA World Cup. In the UK we have our own home-grown headline hero – a black-and-white Collie cross named Pickles. In the country that spawned Sherlock Holmes, Miss Marple and Inspector Morse, we had to rely on a dog to solve one of the greatest mystery thefts of all.

The World Cup was famously hosted and won by England in 1966. Images of the captain, Bobby Moore, holding the trophy aloft while being carried on his teammates' shoulders are highlights of our sporting and cultural history, but only four months before the tournament, the hosts were in the embarrassing position of having lost the World Cup. When I say lost, I don't just mean a match. I mean the World *Cup*.

As a way of building excitement for the eighth tournament, the historic Jules Rimet trophy was brought to London in March 1966. The gold-plated sculpture of Nike, the Greek goddess of victory, was proudly put on display as the centrepiece of the Stanley Gibbons's 'Sport with Stamps' exhibition at Westminster's Central Hall.

However, the second day it was on show, it disappeared from a locked cabinet. Three million pounds' worth of stamps were ignored by the thieves, who clearly only had eyes for the trophy itself. Questions were asked about the security that was meant to have been in place, and everyone panicked. England would be a laughing stock. How could they host the World Cup if they couldn't even keep the trophy safe for two days? The Football Association did what they tend to do. They issued a statement: 'The FA deeply regrets this most unfortunate incident. It inevitably brings discredit to the FA and to this country.' They also commissioned a silversmith to make a speedy replica, despite FIFA expressly forbidding them to do so.

Meanwhile, Joe Mears, chairman of the FA and Chelsea FC, received a phone call from a man who gave the name 'Jackson'. The man told him that a package would be left at Chelsea's Stamford Bridge ground the following day. The package contained the removable lining from the top of the trophy along with a ransom note asking for £15,000. Mears immediately handed it over to the police.

An undercover policeman arranged to meet 'Jackson' in Battersea Park. He carried a suitcase stuffed with newspapers covered with a layer of five-pound notes. 'Jackson' was arrested and revealed to be a former soldier and small-time fraudster named Edward Betchley.

Betchley denied responsibility for the theft, claiming that he was simply a middleman acting on behalf of 'The Pole' (who was never identified, if indeed he ever existed). He was sentenced to two years in prison after being convicted of demanding money with menaces.

The trophy was still missing. Rewards were offered, detectives followed every lead, but the World Cup was nowhere to be found. For seven days the nation held its breath, fearing that the world's greatest football prize would never be seen again.

Then, with the timing of a true performer, Pickles entered the scene. On 27 March in Norwood, South London, Pickles and his owner Dave Corbett stepped out for a Sunday walk. When the dog started to sniff the bushes by a neighbour's car, Corbett went over to investigate and found Pickles fixated by a newspaper parcel tightly bound with string. He took a closer look. He later told journalists:

> I tore a bit off the bottom and there was a blank shield, then there were the words Brazil, West Germany and Uruguay printed. I tore off the other end and it was a lady holding a very shallow dish above her head. I'd seen the pictures of the World Cup in the papers and on TV so my heart started thumping.

He took the parcel to the police station in Gypsy Hill, where he told the duty officer that he had found the World Cup. He was initially met with disbelief: 'Doesn't look very World Cuppy to me, son', said the sergeant. Then he was put under suspicion and found himself being questioned at Scotland Yard. Once the confusion had been cleared up, Corbett returned to Norwood to find the world's press camped out on his doorstep.

Nations hail Pickles, the World Cup hero

Pickles becomes a global media sensation.

Pickles became a global media sensation. He was named 'Dog of the Year', got a silver platter with £53 in cash and was given a year's supply of free food by Spillers. Corbett got a reward that allowed him to put a deposit on a new house.

When England went on to lift the trophy on 30 July, Pickles

and Corbett were invited to the celebration banquet as guests of honour. Perhaps overwhelmed by the attention or overawed at the company, this was the only time Pickles let himself down. He walked over to the lift shaft in the hotel, lifted his leg and did a wee.

After dinner when the players stepped out onto the balcony, Bobby Moore held Pickles up for the crowd to cheer.

Later, Pickles was awarded a silver medal by the National Canine Defence League and starred in a film called *The Spy with a Cold Nose*, alongside June Whitfield and Eric Sykes. He appeared on *Blue Peter* and *Magpie* and received invitations from all over the world. A plaque, erected on Beulah Hill in 2018 in his honour, stands close to the spot where the cup was found. His collar, along with the replica trophy commissioned by the FA, can now be seen at the National Football Museum in Manchester.

After their historic third World Cup victory in 1970, the original cup was awarded permanently to Brazil and a new one (the FIFA World Cup trophy) designed for future tournaments. The Jules Rimet trophy was again stolen, this time from the headquarters of the Brazilian Football Confederation in Rio de Janeiro in 1983. It has never been recovered.

PUDSEY

We have highlighted the heroism of dogs who detect cancer, sniff out bombs, guide those who cannot see or rescue people from collapsed buildings, but in every business there has got to be a bit of show, a sprinkle of glitter, a dash of drama. So what about a dog with the talent to bring an audience to their feet in appreciation?

There is no bigger TV talent programme in the UK than *Britain's Got Talent*. With a cheque for £500,000 and the chance to put on a show in front of the Queen at the Royal Variety Performance, it is a competition worth winning.

Tens of thousands enter every year. The first five winners of the show were all incredible singers or exceptional dance acts. Could series six produce something completely different? Ashleigh Butler, a schoolgirl from Wellingborough, Northamptonshire, certainly thought so, and she was determined to give it her all with her prized co-performer, Pudsey.

Ashleigh, then a sixth former, had been working with her six-year-old Border Collie/Bichon Frise/Chinese Crested cross long before the idea of entering *BGT* even crossed her mind. Once basic puppy training, such as sitting and lying down, had been perfected, she started to teach Pudsey to master simple moves such as waving his paw when sitting, and then moved on to rolling over, weaving through her legs and spinning.

The dog picked up the new skills quickly, adding some more of his own such as jumping through his owner's arms and walking on his back legs. 'These aren't basic moves but they were easy for Pudsey to learn', says Ashleigh, who explained, 'He can jump really high in our routines because he's so used to doing it in "agility", which helps his flexibility and with some of the moves he can do.'

Ashleigh had been taking Pudsey to obedience and agility classes. She then started to research heelwork to music (HTM), using books and online resources. This is a bit like dressage but for dogs – the dog and their human have to dance together, showing a combination of obedience, creativity and energy. Ashleigh learnt the moves along with him, trying out various options to see which ones he liked doing best. 'How long it takes a dog to learn tricks depends on how much time people put in, how much the dog really wants to do it, and whether he's a quick learner', says Ashleigh.

Her application to *Britain's Got Talent* impressed the producers enough to make it through the early stages. Next came the audition in front of the judges. The show had never been bigger nor seen more acts, with auditions taking place in Cardiff, Birmingham, London, Manchester, Blackpool and Edinburgh. The pair walked out in Cardiff to perform in front of Simon Cowell, Alesha Dixon and David Walliams. Their routine, set to the soundtrack from *The Flintstones*, was flawless and, as it came to an end, both audience and judges rose to their feet to give them a standing ovation. Simon, in particular, seemed to admire Pudsey's ability and hinted that he would love to see a dog triumph in the competition.

Their semi-final heat was the first of five, each featuring eight acts. This time they performed to 'Peppy and George'. Again, the dog's dexterity and skill stunned everyone watching and a public vote saw them through to the final which took place on 12 May 2012.

The tune Ashleigh had chosen this time, 'Mission: Impossible', seemed very apt. Could a sixth former and her pet possibly take the title and beat the hot favourites, a singing duo called Jonathan and Charlotte?

Over thirteen million peopled tuned in to the live show as Pudsey and Ashleigh were lowered onto the stage on a suspended chair to begin their high-octane act. Despite the distraction of the crowd, the loud music, the busy set and the giant screen flashing with changing lights, their performance was as perfectly coordinated as ever. Pudsey didn't put a paw wrong as he turned in time with Ashleigh, weaved through her arms, danced on two legs behind her and ran along the judges' desk before leaping onto his owner's back.

As to the judges, Alesha declared, 'For anyone out there who treats animals wrongly, what you're showing them is how special dogs are.' Simon added, 'One of my favourite ever acts – you know how much I love dogs.'

It was all down to the public vote in a nation of dog lovers. Millions were persuaded by Pudsey. As presenters Ant and Dec read out their names, Ashleigh and Pudsey became the first-ever non-human partnership to win *Britain's Got Talent*.

In the wake of their victory, Pudsey-mania swept the land. Invitations poured in from every direction and the Kennel Club was inundated with people asking where they could find classes in agility and heelwork to music.

Ashleigh and Pudsey performed for the Queen twice, at the Royal Variety Performance and at Epsom Downs Racecourse as part of the Diamond Jubilee celebrations – where I first interviewed Ashleigh. She told me that the Queen, a renowned dog lover herself, had taken a particular interest in her training techniques and was captivated by Pudsey's ability to focus on her, ignoring all distractions.

They took part in dozens of television programmes and Pudsey made his acting debut as 'Duchess' in the TV adaptation of David Walliams's book *Mr Stink*. Not to be left out, Simon Cowell produced *Pudsey: The Movie*.

Pudsey's death in 2017 left his owner bereft. She wrote on social media, 'My heart is broken, and I don't know how I'm going to get through this. He was my one in a billion dog that will never be replaced.'

Coincidentally, dog fever also took over the USA in 2012 when the seventh series of *America's Got Talent* was won by Olate Dogs. Another dog called Matisse, with his trainer Jules O'Dwyer, won *Britain's Got Talent* in 2015, and in 2017 ten-year-old Alexa Lauenburger won *Germany's Got Talent* with a team of eight dogs.

RATS

Rats are not top of the popular animal charts. They cause damage in our homes and offices; they carry deadly infections, including salmonella and Weil's disease. Rats were to blame for the bubonic plague, or Black Death, that swept the world in the middle of the fourteenth century. In Central London it is said that you are never more than six feet away from a rat.

Rats get the blame for everything – they'll probably get blamed for the spread of COVID-19 – but let's give the common rat a chance to be a hero rather than a scourge.

Rats are saving lives across the world and putting their own at great risk in the process of detecting landmines. The heroRATS is an army of highly trained African Giant Pouched rats. It was put together by the Belgian not-for-profit organisation, APOPO, to clear landmines in Angola, Mozambique and Zimbabwe, as well as Cambodia and other countries across South East Asia.

Giant pouched rats are indigenous to sub-Saharan Africa. They have a highly tuned sense of smell, allowing them to detect mines buried underground. Their work is more efficient than that of a dog or a human, and giant pouched rats are light-footed enough not to set a mine off if they do happen to step on it.

Rats can also work quickly. It would take a human using a metal detector far longer, with a much higher risk of triggering an explosion if they put a foot wrong. One rat can search an area of 2,000 square feet in twenty minutes. It could take a human up to four days to do the same. Each rat costs several thousand pounds to train. It is a laborious process taking around nine months as the rats get used to working with humans and learn to walk on rope grids in minefields, wearing specially designed harnesses. However, they are cheaper to train than dogs and easier to transport.

Unlike dogs, rats don't respond to verbal commands so they are taught with a system of clicks and rewards of mashed avocado, banana or peanut. When a rat comes across a mine it signals to its handlers by sticking its nose in the air. The mine can then be detonated or deactivated.

It is estimated that there are 110 million landmines still underground in more than sixty countries, some from conflicts that ended decades ago. Landmines do not discriminate between soldier and civilian and pose an enormous risk. Every year, they maim between 15,000 and 20,000 people. Since APOPO's launch in 1997, the rats have helped to clear more than 13,000 mines in Cambodia, Angola, Mozambique and Tanzania.

Sniffing out landmines is not the only super-talent possessed by rats. They have also become experts in detecting a disease that, according to the World Health Organisation, is 'one of the top ten causes of death worldwide'.

The disease is tuberculosis which rats have been trained to detect in sputum. They are used to take a second look at

Molly, the first cat-detection dog. Yes, she finds missing cats rather than chases them.

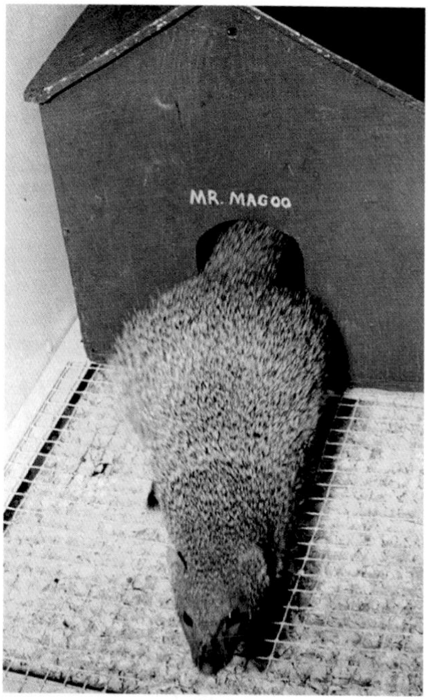

Mr Magoo the mongoose was under sentence of death. His life was saved by the campaign 'No Noose for the Mongoose' which won him thousands of friends across America.

ul the oracle octopus foretold the
tcome of all of Germany's World
ıp matches with pinpoint accuracy
2010.

Popular winners of *Britain's Got Talent* in 2012, Pudsey and Ashleigh Butler meet the Queen after their appearance in the Royal Variety Performance.

An African Giant Pouched rat training to detect landmines sniffs the ground for explosives. They are light-footed enough not to set a mine off if they happen to step on it.

Sadie and her handler Karen Yardley receive their Dickin Medal for saving lives. After a car bomb exploded in Kabul, Sadie picked up the scent of a second device which was defused just in time.

Salty (centre) and Roselle (left), who were both awarded the Dickin Medal for leading their owners to safety from the World Trade Center on 9/11.

Specialist Fire Investigation Dog Sherlock in his fireproof vest and boots. His keen sense of smell is more accurate and faster than technology at determining whether a fire has been started deliberately.

Shrek survived six years in the wild. When shorn live on TV, his fleece weighed sixty pounds – enough to make twenty men's suits.

A murmuration of starlings forms a bird-like shape at dusk in Scotland.

The badly behaved honey badger Stoffel and his Houdini-like talents have attracted visitors from far and wide, helping to raise funds for wildlife rehabilitation.

Stationmaster Tama sitting on the ticket gate in 2008, at the unmanned Kishi Station.

The Tamworth Two: Butch (left) and Sundance, who escaped from an abattoir, are reunited after six days on the run.

Thandi, who survived a brutal attack by poachers, with her calf Thembi in the Kariega Game Reserve in South Africa.

Iris Grace and Thula painting together in the kitchen.

Statue of Matthew Flinders and Trim at Euston Station, London.

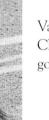

Valegro floats above the ground. He and Charlotte Dujardin celebrate winning their gold medal at the Rio Olympics in 2016.

Alfred Munnings' painting of Major General Jack Seely and Warrior in 1918. They spent four years on the front lines of the First World War together, surviving against all the odds.

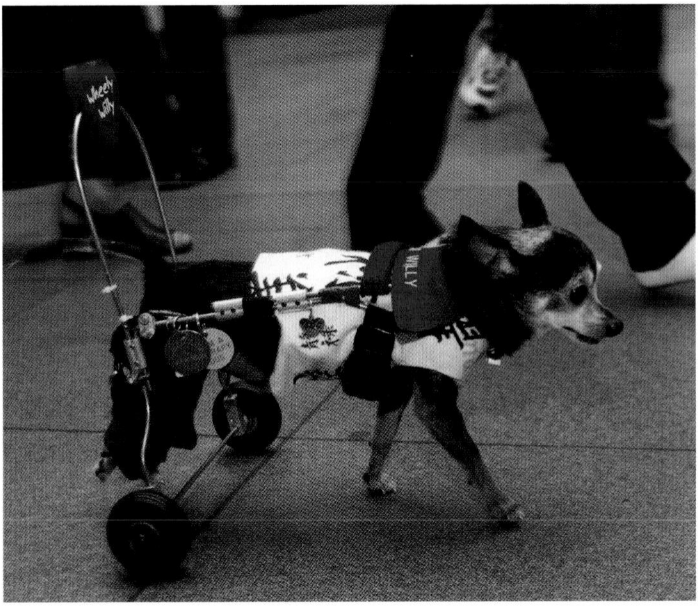

Wheely Willy, a Chihuahua, in a two-wheeled brace supporting his hindquarters, on his first visit to Japan in 2004.

Zarafa, a giraffe, installed in the Jardin des Plantes, Paris. Women wore their hair in topknots to simulate the look of her horns, and her distinctive coat inspired a fashion craze – 'la mode à la girafe' – which swept the country.

samples, thus speeding up the diagnostic process. It would take a lab technician a day to work through, say, forty samples using a microscope, but a rat can do the same work in a matter of minutes.

In 2015, screening over 40,000 samples, rats identified more than 1,000 cases that had been missed by more traditional diagnostic methods. In Tanzania, it's suggested that TB diagnoses have increased by over 40 per cent since the rats started work in 2007. In 2018, research from the Sokoine University of Agriculture in Morogoro, Tanzania, also showed that rats were able to detect up to 70 per cent more cases of TB in children than standard testing. This could be a game changer as tuberculosis affects an estimated one million children a year, with a quarter dying from the disease.

Given their propensity for spreading disease, rodents may be the unlikeliest of heroes, but we should be more respectful of them and grateful for their contribution to the health and safety of the human race.

RECKLESS

Staff Sergeant Reckless has been described as America's greatest war horse. So in tune with her troops was the mare that she would follow them anywhere, eat their food and even drink coffee or beer with them.

Reckless started out as a trainee racehorse in Seoul. She was sold for $250 by a Korean stable boy desperate for money. His sister had stepped on a landmine and needed a prosthetic leg, so he was keen to raise funds. The US Marine Corps wanted a pack horse to carry ammunition to the men on the front line and to rescue the wounded, so the deal was done. Commander Eric Pedersen paid the money and took the horse.

His marines relied on a piece of artillery called the 'recoilless rifle', which could fire a 75mm shell over huge distances with perfect accuracy. However, carrying the 100-pound gun over the battlefield took at least four men and, in tough, bitterly cold weather conditions, it was dangerous work.

Pedersen realised that a horse would make transporting the gun infinitely easier and got his men to train her to carry it, along with more than 200 pounds of ammunition (each 75mm shell weighed twenty-four pounds). As the soldiers had nicknamed the weapon 'Reckless', the horse was given the same name.

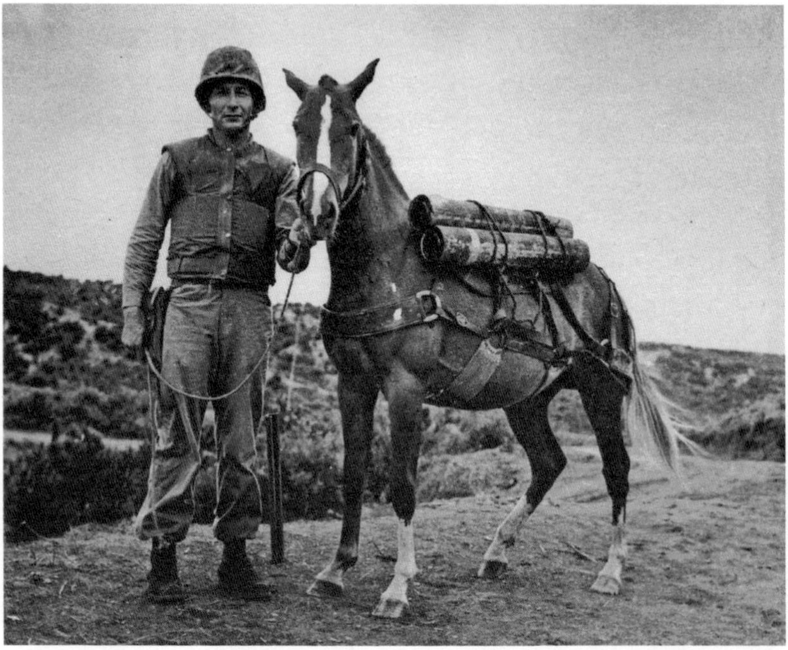

Reckless of the United States Marine Corps loaded with ammunition in Korea.

Before she could head out with the gun, Reckless needed to learn how to survive on the front line. The men taught her how to run for cover or lie down to avoid heavy gunfire. She learnt how to step over barbed wire and to react to certain commands in an instant. She was terrified the first time she heard her namesake being fired. The second time she was calmer, possibly distracted by trying to eat a helmet liner she had found nearby. By the third volley she was positively blasé.

Reckless quickly became more than just a pack horse and the troops forgave her many eccentricities, which included coming inside their tents to sleep alongside them when she was cold, and eating and drinking pretty much everything in sight. Her fondness for Coca-Cola led to her being rationed to two bottles a day by the platoon vet, and she would happily feast on everything from shredded wheat to peanut-butter sandwiches or anything else she could find lying around. This included her own blanket and once, to the fury of the card players, $30-worth of poker chips.

To begin with, one of the men would accompany Reckless on her trips, but eventually, as their numbers were decimated, she went solo. In one day alone, she could haul fresh ammunition to the front lines more than fifty times, through waterlogged paddy fields and across steep and treacherous terrain into the 'smoking, death-pocked rubble' of the battlefield. She was undaunted by the heavy shelling and the day she was hit twice by shrapnel, receiving wounds above one eye and in her left flank, she travelled more than thirty-six miles and carried 9,000 pounds of shells. She also took a number of the wounded to safety. Every step of the way she put her comrades first.

She served in many battles and despite the dangers never

hesitated in her duties, be they lugging heavy artillery or carrying the wounded. Even at the Battle for Outpost Vegas, where 1,000 American and 2,000 Chinese soldiers were killed in just three days, she remained steadfast as the carnage raged around her.

She was promoted to sergeant and then staff sergeant, and her numerous awards, which she wore pinned to her blanket, included two Purple Hearts, a Marine Corps Good Conduct Medal, the National Defense Service Medal, the United Nations Korea Medal, a Korean Service Medal, a Navy Unit Commendation, a Republic of Korea Presidential Unit Citation and a Presidential Unit Citation with bronze star. She inspired awe and respect in everyone who saw her or heard about her bravery.

After the war, the public lobbied for her to be brought back to America with the troops she had served alongside in the East. Pacific Transport Lines offered her free passage on a ship to San Francisco, and she arrived there in November 1954, just in time to attend the Marine Corps Birthday Ball. There, she was offered cake – which of course she ate, along with the flower arrangements.

RED RUM

The 1970s was the decade of glam rock, punk and platform shoes. Kids rode Choppers, ate Spangles and bounced on space hoppers. Grown-ups drove Ford Cortinas while decked out in tank tops and flares. We had strikes, three-day weeks, a winter of discontent ('79) and a summer ('76) so hot that the roads melted. It was a decade where things were changing fast but the one certainty, the one constant, the one thing that was absolutely regular was that, every April at Aintree, there would be Red Rum.

For five years running he ran in the Grand National and for five years running he finished in the first two, winning the great race three times, finishing second twice and jumping 150 fences without a mistake. He was the most famous horse in the land, and everyone loved his unlikely story.

Born in Co. Kilkenny, Ireland, he was named using the last three letters of his sire and dam (Mared and Quorum): Red Rum. He started out as a sprinter and dead heated to win his first race (ironically at Aintree, which used to stage a mixed jumping and flat meeting). He was ridden twice by the great Lester Piggott.

Red Rum was not a big horse but he was considered to have jumping potential by Bobby Renton, who had trained Freebooter to win the Grand National in 1950. He had a mixed couple of seasons with a few wins and plenty of hard races. The major concern was that he was often lame. He was diagnosed with pedalostitis, an inflammatory condition affecting the main supporting bone in the foot and causing pain and tenderness. After a long losing run he was sent to the Doncaster sales in 1972.

Meanwhile in Southport, twenty miles north of Liverpool, a former taxi driver and used-car salesman turned racehorse trainer called Donald McCain (known as Ginger) had a wild dream of trying to train a horse to win the Grand National. One of his former passengers was a local businessman who had started out as a trawlerman and was now a millionaire. On Saturday evenings when Ginger drove Noel le Mare to and from the Prince of Wales Hotel in Southport, they would talk racing. Eventually, McCain persuaded Le Mare to send him a horse to train.

Not knowing about his issues with soundness, they paid 6,000 guineas for Red Rum and he was stabled in the yard behind McCain's car showroom. Most racehorse trainers have access to grass or all-weather gallops. Some have covered rides or indoor schools, mechanical walkers and equine swimming pools. McCain didn't have any of these, but he did have the endless miles of Ainsdale beach and the Irish Sea.

McCain sent Red Rum through the streets of Southport down to the beach and watched him trot out on the sand. To his horror, he could see the horse he had just bought for more money than he'd ever paid for a racehorse, was lame. 'Oh my God, what was I going to say to Noel le Mare whose money I had spent?' he thought.

He sent him into the sea for a paddle. Cold salt water has long been used to treat lameness and, in this case, it was a miracle cure. Red Rum came out of the sea and trotted out perfectly sound. He cantered on the sands and he was fine. Only months later did McCain find out about the pedalostitis and, by then, he had already found the answer.

Within seven months 'Rummy' had won six of his nine races, with winnings of almost £30,000. McCain's dream of having a runner with a chance in the Grand National was on track. Red Rum was sent off, the 9/1 joint favourite with the Australian champion Crisp.

The race itself was in the doldrums. Aintree Racecourse was up for sale and under threat of being turned into a factory or a housing development. The crowd was dwindling year by year as the prices for tickets went up, and the newspapers predicted every time 'This could be the last Grand National'. The race needed a hero.

In 1973 it looked as if that hero would be Crisp, a giant of a horse who jumped like a buck and whose huge stride propelled him into a massive lead. Red Rum, ridden by Brian Fletcher, was popping away over the fences but was miles behind. Crisp was a good 100 yards (nearly half a furlong) clear at one stage, but as he jumped the last fence it was obvious the big horse was tiring. His jockey Richard Pitman resorted to his whip but the impact threw Crisp off balance and he was suddenly galloping up and down on the spot.

Behind him, having negotiated the last fence, Red Rum was making ground. As Crisp got more and more tired and the weight of twelve stone on his back took its toll, so Rummy, carrying twenty-three pounds less in weight, kept galloping. Just before the winning post, Red Rum took the lead and won his first National by three-quarters of a length. Both horses had smashed Golden Miller's record time and the new mark of nine minutes, 1.9 seconds would stand for another seventeen years. No one could quite believe what they had just witnessed, not least Richard Pitman, who had been so sure he was about to win. 'I felt as though I was tied to a railway line with an express train thundering up and being unable to jump out of the way', he said.

Having been the villain of the piece, Red Rum had to shoulder top weight of twelve stone in 1974. He was ridden patiently by Brian Fletcher, sidestepping fallers and making his way towards the front by the time they reached Becher's Brook on the second circuit. He was then sent into a lead that he never relinquished, winning by seven lengths from the dual Cheltenham Gold Cup winner, L'Escargot. Red Rum became the first horse for thirty-eight years to win consecutive Grand Nationals and the first to carry twelve stone to victory. This time, the crowd appreciated his victory. A few weeks later he also carried top weight and won the Scottish Grand National.

In the 1975 race, on soft ground, Red Rum was second to L'Escargot. In 1976, with a new jockey on board – Tommy Stack – he was beaten by Rag Trade, who had twelve pounds less on his back. Red Rum's form seemed to be fading, but McCain kept the faith and he lined up in 1977 once again, as top weight, against forty-one other runners. At the age of twelve, many people thought he was over the hill, but the general public were behind him and he was again sent off joint favourite.

The crowds at Aintree, which had dwindled in the early '70s, were boosted to 51,000 coming out in the sunshine to watch Red Rum. It was one of the greatest moments in horse-racing history, as they cheered him on to win by twenty-five lengths from Churchtown Boy. Peter O'Sullevan was commentating:

> The crowd are willing him home now. The twelve-year-old Red Rum, being preceded only by loose horses, being chased by Churchtown Boy . . . They're coming to the elbow, just a furlong now between Red Rum and his third Grand National triumph! It's hats off and a tremendous reception, you've never heard one like it at Liverpool – Red Rum wins the National!

O'Sullevan had no doubt of the role in British racing played by the legendary horse. 'McCain and Red Rum appeared instrumental in saving the Grand National at a time when it was very much in peril', he wrote.

Red Rum wins his third Grand National at Aintree in 1977.

The first horse ever to win the Grand National three times, Red Rum was propelled into public folklore. He was celebrated in the ballroom of the Bold Hotel in Southport (which had laid out red carpet specially) and stole the show with a live appearance in the BBC studios during *Sports Personality of the Year*.

He was due to return to Aintree to try to win for a fourth time in 1978 but a hairline fracture meant he was retired from racing. The injury was the lead story on the *Nine O'Clock News* and made all the front pages. Instead of running in the race, Red Rum led the Grand National parade and continued to do so for another fifteen years.

Red Rum was a celebrity in his own right, making personal appearances for fees comparable to those earned by the comedians and entertainers of the time. He switched on the Blackpool illuminations, opened the new Steeplechase ride at the Pleasure Beach, appeared on several TV shows and opened various supermarkets.

The horse who never fell in over a hundred races over jumps died in October 1995 at the ripe old age of thirty. He was buried at the winning post of Aintree. His gravestone carries the epitaph:

Respect this place
this hallowed ground
a legend here
his rest has found
his feet would fly
our spirits soar
he earned our love
for evermore

RIFLEMAN KHAN

In early 1942, the government issued a call for dogs to help with the war effort. Thousands responded to the request for strong, fit, intelligent dogs to be trained for rescue work and for guard and patrol duty with the army. Among them was eight-year-old Barry Railton who volunteered Khan, his handsome German Shepherd.

As War Dog 147, he showed promise from the moment he arrived at the War Dog Training School. His intelligence and skill marked him out as a star pupil when it came to detecting explosives. The school had been recently established at the greyhound racing kennels in Potters Bar, and by May 1944 76,000 dogs had graduated through its ranks. Eighteen of them would go on to win the PDSA Dickin Medal, the canine equivalent of the Victoria Cross.

The German Shepherd was assigned to the 6th Battalion Cameronians (Scottish Rifles) and became known as Khan after the Indian soldiers serving with the Allied forces. His handler was Lance Corporal James Muldoon, known as Jimmy. The pair worked well together and quickly formed a close bond.

The battalion was despatched to Belgium to fight in the Battle of the Scheldt in November 1944. Khan and Muldoon were among the troops attacking the island of Walcheren. It was a crucial mission because recapturing the island would allow Allied forces to begin their invasion of Germany.

Approaching the island under cover of night, their unit was caught in an enemy spotlight. The Germans started shelling them with heavy artillery fire. The assault craft capsized, throwing the soldiers into the icy water. Khan managed to get to shore but his handler was nowhere to be seen. Muldoon was in big trouble: he couldn't swim and was wearing a heavy pack which dragged him down beneath the waves. Even if he managed to avoid being hit by enemy fire, he would surely drown.

But Khan braved the heavy shelling to swim out 200 yards from land and drag his handler back to the shore. He refused to leave Muldoon's side and even accompanied him to hospital.

The other soldiers from the unit who witnessed Khan's rescue later called for the dog to be decorated for his loyalty and bravery. This was informally recognised with a promotion to 'Rifleman Khan', and more formally with the award of a Dickin Medal in March 1945. The citation read: 'For rescuing L/Cpl Muldoon from drowning under heavy shell fire at the assault of Walcheren, November 1944, while serving with the 6th Cameronians (SR).'

James Muldoon and his devoted rescuer Rifleman Khan.

After the war, Khan returned to the Railton family. In July 1947, along with other dogs who had received the Dickin Medal, he was invited to take part in a parade at the National Dog Tournament at Wembley Stadium. Barry wrote to Muldoon to invite him along and the soldier travelled down from Scotland to see his old pal.

When the German Shepherd realised Jimmy was in the arena, his joy was irrepressible. He nearly knocked Muldoon to the ground, jumping up at him over and over again. It was the first time for two years that the two had seen each other, and the strength of their bond was clear to everyone.

It was so strong that Barry couldn't ignore it. He still thought of Khan as his dog, and loved him, but he could see that Khan and Muldoon were meant to be together. So he and his family handed their beloved dog over to the man whose life he had saved, and the two best friends went back to Scotland to spend their remaining years living happily together.

In 2019, Avondale councillor Margaret Cooper launched a campaign to raise £55,000 to erect a bronze statue of the soldier and his heroic dog in Strathaven to keep the memory of Khan's rescue alive for future generations.

RIP

Many lives were saved by the heroism of dogs during the Second World War. Not just those of the soldiers on the front line and on battlefields, but also those of civilians in the cities of Europe.

The very first civil defence search-and-rescue dog was Rip. Unlike the many others who were to follow, he had no formal training. It all happened by accident, but it was Rip's outstanding work that paved the way for dogs to be used to search for survivors after a bombing raid.

After a heavy raid on London in 1940, a wire-haired terrier was found wandering the streets, picking his way through the rubble. Air Raid Warden King came across him in the debris and threw him some scraps from his lunch.

Rip ready for action amongst rubble and debris following an air raid in Poplar in 1941.

There was something really engaging about this particular dog. Mr King took him back to his base, the Southill Street Air Raid Precautions Station, and Rip was adopted as their mascot. Heading out with his colleagues after every raid, Rip soon began to act as an unofficial rescue dog, taking to the task instinctively. Perhaps when he saw people clambering over the ruins looking for survivors he thought it was a game. He joined in and turned out to be better at it than anyone. King said, 'It wasn't a question of training him. We simply couldn't stop him.'

There's no such thing as a simple smell to a dog. They can read the clues of the world in a way that we just can't. A human nose has around six million smell-detecting cells while a dog's cold, wet nose has nearly 300 million.

After a particularly heavy night of bombing in the East End, Mr King searched through the remains of what had once been a residential street. It was now nothing more than smoking rubble. Rip stood still for a moment, nose twitching, before clambering over the shattered masonry to a smouldering pile of bricks. He began to paw frantically at the debris, barking loudly to attract the attention of the wardens who came over to investigate. The men dug deep into the remains of the ruined building and Rip barked excitedly when they uncovered an unconscious child and carried him to safety.

This was just one of more than a hundred lives saved by Rip during the Luftwaffe air raids in a twelve-month period between 1940 and 1941.

The risks he faced were enormous. There were often fires burning inside the buildings and sometimes unexploded bombs. The walls were unstable and there was broken glass everywhere, but once Rip had picked up a scent, he refused to leave the site until he'd traced it. Rip's work earned him a Dickin Medal 'for locating many air-raid victims during the blitz of 1940'. He proudly wore the medal on his collar for the rest of his life. It was later sold at auction for the record price of £24,250 in 2009.

ROCKY

When a serial burglar was foiled mid-theft by a bird, the creature's detective work earned him not only plaudits from police, but a new nickname: Hercule Parrot.

Septuagenarians Peter and Trudy Rowing's home in Kent

was broken into late one night in June 2017. The thief snatched a laptop, a phone and two oxygen canisters Trudy needed for breathing problems. Then he spotted Rocky, the African Grey parrot, and decided that he might be worth a bob or two as well.

He tried to take Rocky out of his cage, but the parrot was having none of it and bit thirty-seven-year-old Vitalij Kiseliov on the hand. He bit deep and blood poured from the wound as Kiseliov made a run for it. When police arrived at the scene, it offered them the vital evidence that would bring the serial thief to justice.

Rocky's bloodletting bite led to the arrest and conviction of the serial burglar.

DNA testing to identify a suspect using their unique genetic blueprint can make all the difference when it comes to solving crime. It was first used in a criminal investigation in 1985 and has become ever more important, both in nailing offenders and exonerating the innocent.

Leaving fingerprints can be avoided by wearing gloves, but tiny samples of biological evidence – drops of blood or sweat, a single hair, flecks of dry skin, earwax – can all be left unwittingly and used to identify a suspect.

Kiseliov was already on the police database because of former misdemeanours so, after DNA testing on samples of the blood, it was relatively easy for him to be found and arrested. At Maidstone Crown Court he admitted to six counts of burglary and was jailed for four years.

As for Rocky, who went missing after being thrown out of the window by the thief, he was found and returned to the Rowings after a Facebook appeal by their granddaughter Nikki.

SADIE

When a suicide bomber caused devastation in Kabul on 14 November 2005, it was Sadie, a black Labrador with soft brown eyes, who set to work sniffing out any further explosives.

Sadie was nine and so keen on her job that she was still working, despite being past the usual retirement age. She had been trained by the Royal Army Veterinary Corps in Leicestershire and was part of the 102 Military Working Dog Support Unit. Together with her handler Lance Corporal Karen Yardley, she had completed two tours of duty, serving in Bosnia, Iraq and Afghanistan, and was now part of the NATO International Security Assistance Force in the Afghan capital.

The suicide car bomb exploded outside the UN headquarters in Kabul, killing one ISAF soldier and injuring seven others. There was mass panic as troops arrived to try and calm the situation and medics tended the wounded. Second bombs are common after attacks of this type, planted to maximise casualties by also killing those who rush to help. Lance Corporal Yardley and her dog went straight to work and Sadie picked up a scent by the compound's external wall. When she sat and stared at it, Yardley knew that she was on to something and shouted for everyone to get as far away as possible.

Sure enough, behind the two-foot-thick concrete was a pressure cooker filled with TNT. It had been hidden under sandbags but contained enough explosives to kill or maim everyone working in the area. As it was, bomb disposal experts were able to defuse the device just in time.

Both Lance Corporal Yardley and her dog were recognised for their bravery, and Sadie was presented with the Dickin Medal for 'outstanding gallantry and devotion to duty' by HRH Princess Alexandra in a ceremony at the Imperial War Museum. The citation concluded that 'Sadie's actions undoubtedly saved the lives of many civilians and soldiers.'

Yardley said of Sadie, 'She's my best friend. We're as close as you could possibly be. And when Sadie retires, my mum even wants to adopt her.'

Now that she has retired, that is exactly what has happened, and Sadie lives happily with the Yardley family in Scotland.

SALTY AND ROSELLE

Earlier in *Heroic Animals* I paid tribute to Jake and the other dogs who worked on finding survivors of the attack on the World Trade Center on 9/11. It turns out that they were not the only dogs who saved lives that day.

At the time of the attacks, Omar Rivera was working at the Port Authority of New York and New Jersey headquarters on the seventy-first floor of the North Tower.

As the American Airlines Flight 11 hit the building between the ninety-third and ninety-ninth storeys, Rivera was at his desk. He heard a deafening booming noise as the building started to sway. His computer crashed to the ground and when he smelt smoke, he realised that something was terribly wrong.

Forty-three-year-old Rivera had gone blind fourteen years earlier. But with the help of his guide dog Salty, he had been able to continue working as a senior systems designer and to travel through Manhattan using the subway.

Salty was a yellow Labrador, who was born in 1996 and trained as a guide dog by Guiding Eyes for the Blind in 1998. Rivera and Salty were introduced five months later. It was a perfect match. 'Trust is the most important thing in a relationship', Rivera has explained. As a pair, they had complete trust in each other and, on that day, it saved his life.

Salty had been anxious as soon as the first crash was heard. He was trying to convey to Rivera that they needed to move immediately. He led his owner to the crowded central stairwell. It was chaotic. There was debris everywhere and people were panicking, rushing to leave the building as quickly as possible. There was no space for a man with a guide dog. Rivera explained in a *National Geographic* documentary, 'It was probably too much for him so I said, "maybe Salty, it's better for you to go".' Rivera dropped his harness to let the dog go, so that he would have a chance of making his way out of the building alive. He describes how Salty then decided, '"No, I cannot go without him", so he came back. He was telling me "I am with you, no matter what. I am with you."'

For over an hour, Salty guided Rivera down to the lobby. Once they had managed to get through the doors they ran for their lives and they were only just in time. They were no more than three blocks away when they heard the tower fall. Salty had saved Rivera from almost certain death.

Computer sales manager Michael Hingson was in the same situation, working on the seventy-eighth floor. Hingson had been blind from birth and, like Rivera, relied heavily on his guide dog to take him to the office every day.

Roselle, another yellow Lab, was born in 1998 and introduced to Hingson a year later by Guide Dogs for the Blind. She was asleep under his desk when the plane hit the tower eighteen storeys above them. Despite the noise and chaos around her, she led Hingson and his colleagues to staircase B, helping them into a darkened stairwell where it took just over an hour for them to descend the 1,463 steps.

As they left, the South Tower collapsed, showering them both with rubble. Roselle continued to guide her master away from the scene, walking through the panicking crowds and into a subway station. She took the whole thing in her stride. Hingson said, 'She saved my life. While debris fell around us and even hit us, Roselle stayed calm. While everyone ran in panic, Roselle remained totally focussed on her job.'

In 2002, Salty and Roselle were both awarded the Dickin Medal, only the second time the award has ever been made jointly (the first was to Punch and Judy, a pair of Boxer dogs in 1946, for saving two British officers from an armed intruder in Jerusalem).

The citation for Salty and Roselle read: 'For remaining loyally at the side of their blind owners, courageously leading them down more than 70 floors of the World Trade Center and to a place of safety following the terrorist attack on New York on September 11, 2001.' Both dogs were also given a 'Partners in Courage' award by Guiding Eyes for the Blind and were recognised by the Guide Dogs for the Blind Association.

Roselle did more than save Hingson's life, she changed it. The events of 9/11 inspired him to write a book about their experience – *Thunder Dog: The True Story of a Blind Man, His Guide Dog, and the Triumph of Trust at Ground Zero* – and he went on to become the public affairs director of Guide Dogs for the Blind, the organisation which had brought them together.

SECRETARIAT

Known as 'Big Red', he raised the nation's psyche at one of the most difficult times in its history and became known as America's greatest racehorse – but what was it about

Secretariat that left everyone who watched him race almost unable to believe their eyes?

He was born at Meadow stud in Doswell, Virginia, in 1970. His nickname was an obvious one: he was a red chestnut (with three white socks and a star) and, at sixteen hands two inches when fully grown, his girth was so deep and large that he needed specially made straps to reach round his enormous seventy-six inches. His stride was also huge, at twenty-four feet, eleven inches.

It's fair to say that his early reviews were a mixed bag. One exercise rider, Charlie Davis, called him a 'big fat sucker', though jockey Jim Gaffney described his first ride in 1972 as 'having this big red machine under me, and from that first day I knew he had a power of strength that I have never felt before'.

In his first race he finished fourth, but then went on to win seven of his next eight. Then in August 1972, he ran in the Sanford Stakes, a six-furlong sprint held every year at Saratoga, New York. Secretariat was regarded as second best to Linda's Chief, the odds-on favourite, but powered through the horses ahead of him, 'like a hawk scattering a barnyard of chickens', according to sportswriter Charles Hatton in the *American Racing Manual.* Hatton had already nailed his colours to the mast, writing about Secretariat in an earlier race:

> I never saw perfection before. I absolutely could not fault him in any way. And neither could the rest of them and that was the amazing thing about it. The body and the head and the eye and the general attitude. It was just incredible. I couldn't believe my eyes, frankly.

The superlatives were rolled out with every race Big Red took part in that season, which also saw him win the Eclipse Award as Horse of the Year, the first two-year-old ever to be ranked above his elders for the award. He was syndicated for a record-breaking $6.08 million to help get his owner Penny Chenery out of a sticky financial situation. It looked as though those who bought one of the thirty-two breeding shares had invested well because, as a three-year-old, Secretariat was unstoppable.

In 1973, he became the first horse for twenty-five years to win all three races that make up the US Triple Crown: the Kentucky Derby, the Preakness and the Belmont Stakes. He set record times in all three. His winning time for the Kentucky Derby still stands to this day. In front of a crowd of 134,476 – then the largest racing crowd ever in American history – at Churchill Downs, he became the first horse to win in under two minutes.

Ron Turcotte and Secretariat winning the Kentucky Derby in record time in 1973.

In the build-up to the Belmont Stakes at Belmont Park, New York, Secretariat appeared on the front cover of *Sports Illustrated*, *Time* and *Newsweek* magazines. He was in such demand that the William Morris Agency was employed to oversee his public relations. Only four other horses lined up to take him on, and his jockey Ron Turcotte was so convinced he would win that he said he would retire on the spot if he got beaten. 'Riding him was like flying a fighter jet compared to an ordinary aeroplane', he said.

The Belmont of 1973 was watched by over fifteen million people on TV and none of them could believe what they were witnessing. Secretariat and Sham, who had finished second to him in the Kentucky Derby, set off at breakneck speed. They ran the first six furlongs in a faster time than most sprinters can manage, and they still had another six to run. Sham couldn't keep it up but Secretariat kept going faster. Ron Turcotte knew he was due to have a break after the Belmont, so he took the handbrake off and let him go as fast as he wanted. The huge crowd clapped from before the turn into the home straight as he opened up a bigger and bigger lead. Secretariat won by thirty-one lengths in a world-record time for a mile and a half: two minutes, twenty-four seconds. Experts shook their heads in disbelief.

William Nack of *Sports Illustrated* wrote, 'Secretariat suddenly transcended horse racing and became a cultural phenomenon, a sort of undeclared national holiday from the tortures of Watergate and the Vietnam War.'

Secretariat took his new-found fame in his giant stride, even learning to pose for the many cameras pointed in his direction. He was intelligent and kind, gentle and patient. All in all, he deserved the fan mail that came pouring into his personal secretary (yes, honestly). 'The public just couldn't get enough of him. He brought racing back to where it was in the golden era', Turcotte said. Secretariat was still getting fan mail nearly fifty years after his heyday.

The plaudits continued to flow and for the second year running he was named Horse of the Year. But America's icon was at the end of his exceptional racing career. The big money syndication had come with a condition: at the end of his third year, Secretariat would be sent to stud to make mega bucks as a stallion. That was when investors would be able to get their money back. To their horror, initial reports suggested he might have a problem with his fertility, and he sired a relatively small crop of only twenty-eight foals in his first season at stud. Eventually Big Red went on to sire 663 foals, several of whom proved to be successful racehorses in their own right: they included 341 winners and fifty-four stake winners.

His death in 1989 was mourned by millions and the autopsy revealed the secret to his success. His heart was a massive twenty-two pounds in weight, far bigger than the usual size for a horse. Dr Thomas Swerczek said,

> I've seen and done thousands of autopsies on horses, and nothing I'd ever seen compared to it. The heart of the average horse weighs about nine pounds. This was almost twice the average size, and a third larger than any equine heart I'd ever seen. And it wasn't pathologically enlarged. All the chambers and the valves were normal. It was just larger. I think it told us why he was able to do what he did.

SEFTON

At 10.43am on 20 July 1982, sixteen members of the Blues and Royals were making their way through Hyde Park to take part in the daily Changing of the Guard ceremony at Buckingham Palace. As they passed through South Carriage Drive, a huge bomb planted in a nearby car exploded, leaving four men and seven horses dead. Less than two hours later, a second explosion in Regent's Park killed seven bandsmen of the Royal Green Jackets as they played music from *Oliver* to 120 spectators.

It's impossible for anyone who watched the news on that terrible day to forget the scenes of devastation after the IRA bombs were detonated. So many lives were lost and none of the regiment's horses escaped unscathed. The most critically injured of them all was Sefton.

Sefton had joined the army in 1967, moving from Co. Waterford, Ireland, to London, to become part of the Household Cavalry Mounted Regiment based at the Wellington Barracks in Birdcage Walk. He was half Irish Draught, half thoroughbred and had a reputation of being a bit of a character. He would stop suddenly and refuse to budge (particularly going away from the stables), he would fidget and break rank. He was known by the nickname of Sharkey because he had a habit of biting soldiers or horses he didn't like. With his white blaze and four white socks, he stood out from the crowd.

In 1969 he was sent to Germany with the Blues and Royals where less formal duties suited him well. His speed and bravery made him hugely popular at the Weser Vale Hunt. He also won a point-to-point and excelled at show jumping, being part of the British Army of the Rhine team.

Sefton returned to London as emergency backup after an outbreak of strangles (a dangerously contagious respiratory disease) left the Knightsbridge Barracks short of large black horses to perform key ceremonial duties. Over the next four years he undertook Household Cavalry guard duties such as Trooping the Colour. He also took part in various show-jumping competitions until he reached his eighteenth birthday in 1980, when it was time to wind down a bit and concentrate on the main job.

He was nearly twenty when the nail bomb almost destroyed his life. Of the nine horses caught up in the bombing, Sefton was the worst hit of those who survived. He somehow remained standing, and it was only when his rider dismounted that he discovered the full extent of the horse's injuries.

His left eye had been badly damaged. A six-inch nail had gone through his bridle. He had thirty-four wounds on his body and twenty-eight pieces of shrapnel remained embedded deep in his flesh. And, worst of all, his jugular vein had been severed. Sefton was rapidly bleeding to death.

The sound of the explosion had brought everyone from the barracks running to help. Regimental Commander Lieutenant Colonel Andrew Parker-Bowles took charge, initially ordering a soldier to remove his shirt to try and staunch the blood flow from the wound in Sefton's neck. But Major Noel Carding, the Household Cavalry's veterinary officer, realised that the only chance of saving Sefton was to operate as quickly as possible.

They managed to get him into the first available horsebox and back to barracks and into emergency surgery which lasted more than eight hours. Sefton's injuries, the first war wounds suffered by a British cavalry horse for more than half a century, were grave. The shrapnel had gone so deep that in some places it had embedded itself in bone. He lost a huge amount of blood and his body was traumatised by shock. Vets gave him a 50/50 chance of pulling through.

Sefton is shown cards and donations left by well-wishers at Knightsbridge Barracks in July 1982.

Sefton's character and personality came to the fore. He was a fighter and slowly his health began to improve. His progress was followed by well-wishers across the globe. Hundreds of cards and packets of mints arrived at the hospital. Donations of more than £620,000 were used to build a new wing – the Sefton Surgical Wing – at the Royal Veterinary College.

Sefton's extraordinary recovery made him a beacon of hope, something that was intensified when he returned to active service only months after the bombing. While other horses caught up in the attack remained nervous, jumping at the slightest sound, he showed no sign of panic, even when he passed the spot where he had almost lost his life. Sefton was a symbol of resilience and courage that so many people found inspiring.

When he was named Horse of the Year, he received a standing ovation. In August 1984 he was retired from the Household Cavalry and lived out his retirement at the Home of Rest for Horses where he died at the age of thirty.

SERGEANT BILL

The story goes that it was when a wild goat walked into the battle at Bunker Hill near Boston during the American War of Independence and led the Royal Welch Fusiliers from the field, their tradition began of using goats as military mascots. But it was already long established even then: a document written in 1771 declares that 'The Royal Regiment of Welch Fusiliers has a privilege and honour of passing in review preceded by a Goat with gilded horns . . . [and that] the corps values itself much on the ancientness of the custom.' In 1884 it got the royal seal of approval when Queen Victoria presented the regiment with a Kashmir from her own royal herd, and since then there has been an unbroken series of goats sent by the monarch to front the Fusiliers.

All Royal Welsh* goats hold a rank within the regiment, though their records are not always illustrious. Lance Corporal

* The regiment changed their name to the rather more straightforward Royal Welsh on 1 March 2006 when they amalgamated with the other line Welsh infantry regiment.

William 'Billy' Windsor I, who served in the 1st Battalion between 2001 and 2009, spent three months demoted to fusilier in 2006 following inappropriate behaviour. While on active duty in Cyprus, he was part of the parade held to celebrate the Queen's eightieth birthday. Big Bad Billy was having a disappointing day. He refused to keep in line, failed to keep in step and headbutted a drummer. He was charged with 'lack of decorum', 'disobeying a direct order' and 'unacceptable behaviour'. Three months later, thanks to much better behaviour on the parade ground, he was promoted again to Lance Corporal and regained his right to use the corporals' mess.

In 2018, the 3rd Battalion's newest mascot, Shenkin IV, gave his new regiment the slip and spent four weeks in the dense shrubbery of the Great Orme, a limestone headland near Llandudno. He was eventually retrieved by park wardens and an RSPCA vet, and admitted for six months' training at Maindy Barracks in Cardiff.

The Welsh are not alone in revering the regimental goat.

At the start of the First World War, the newly formed 5th Battalion Canadian Mounted Rifles, an infantry unit of the Canadian Expeditionary Force (CEF) who became known as 'The Fighting Fifth', made its way to the Valcartier training camp north of Quebec City.

As the 5th travelled to the camp by train, they passed through a small town called Broadview where they spotted a goat grazing, looked after by a young girl named Daisy. She agreed to their request to take the goat with them 'for luck' and so the goat boarded the train to become 'Sergeant Bill'.

Sergeant Bill in his regimental uniform.

Mascots were supposed to be ornamental and were kept away from front-line duty, but when the Canadian troops were shipped out from Quebec to spend the winter undergoing further training on Salisbury Plain, they refused to leave their lucky goat behind. Bill was bound for Britain. When they later set out for the front, Sergeant Harold Baldwin wrote in *Holding the Line* (1919),

> We could not part with Billy; the boys argued that we could easily get another colonel, but it was too far to the Rocky Mountains to get another goat. The difficulty was solved by buying a huge crate of oranges from a woman who was doing brisk trade with the boys. The oranges sold like hot cakes and in a jiffy the orange box was converted into a crate and Billy [was] shanghaied into the crate and smuggled aboard the train.

While stationed in France, Bill twice found himself put under military arrest, once for eating important documents (including the battalion personnel roll), and a second time for charging at a superior officer. At that point questions were asked whether the goat might be a traitor or an enemy spy.

Bill soon redeemed himself and became something of a hero. At Neuve Chapelle in February 1915, he headbutted three of his comrades into a trench just seconds before a shell exploded exactly where they'd been standing. They all owed their lives to the goat, who was promoted to sergeant for his bravery and quick thinking. At Ypres, where he sustained a number of shrapnel wounds, he was found in a shell crater, resolutely standing over a Prussian guardsman whom the Canadians were then able to take captive.

During the Second Battle of Ypres, Bill disappeared after being gassed by enemy forces. It was feared he had fallen into the hands of the Bengal Lancers who were used to eating a lot of goat curry. He later returned, thankfully in one piece.

During the course of the war he got trench foot at Hill 63 in 1915, shell shock during the battle for Vimy Ridge in 1917 and sustained a number of injuries caused by shrapnel. Any of these might have taken out a lesser goat, but Bill survived them all and never let them keep him from serving his country.

At the end of the war, he was the only original mascot to enter Mons on Armistice Day and one of the few remaining soldiers of the 5th left in active service. With the others he headed to Berlin, where he marched proudly in the Victory Europe parade wearing 'full dress blues with sergeant chevrons and wound stripes "for wounds received and services rendered"'. His actions also earned him the Mons Star, a General Service Medal and a Victory Medal. He returned to Canada a hero and led the parade for his regiment. After his retirement, he was reunited with his original owner, Daisy.

SEVASTOPOL TOM

It was 1855, towards the end of the Crimean War and after a year-long siege, when British and French troops finally managed to capture the Russian port of Sevastopol, the home of the Tsar's Black Sea fleet. They had suffered heavy losses and those who remained were hungry and exhausted, their supplies long gone. The British troops scoured the city for anything edible that might have been left by the Russians but could find no food.

Things were looking desperate when they spotted a small and playful tabby kitten sitting between two of their wounded colleagues atop a large pile of rubbish. They called him Tom and puzzled over how the cat seemed so healthy and well fed, while everyone around him was struggling to find anything at all to eat.

They decided to follow the kitten on one of his sorties among the ruins of the city. When he disappeared under a pile of rubble and didn't return, they dug through the debris to find him. And find him they did – inside a storeroom stuffed full of provisions that the Russians had hidden away at the start of the siege and which had also kept the cat going ever since.

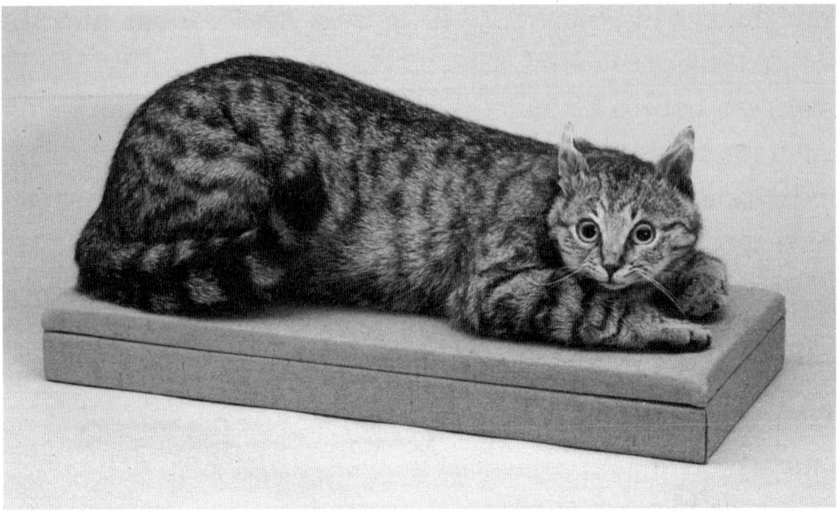

A stuffed Sevastopol Tom.

Thanks to Tom, British and French forces were saved from starvation. They dubbed him Sevastopol (or Crimean) Tom, and when they headed back to Europe, a soldier named William Gair took Tom along too.

SHERLOCK

The bravery and pluck of the men and women who serve with the London Fire Brigade (LFB) cannot be underestimated, but since 2000, four-legged members of the brigade have also demonstrated great courage and dedication, time and again. Of the three dogs currently serving, the aptly named Sherlock has been there longest, and has proved himself a hero on many occasions.

Fire dogs were first used in the USA in the 1980s. Having

seen the benefits they were able to bring to reviews, British fire investigator Clive Gregory began to look into how they might be used over here. The UK's first fire dog, a black Labrador called Star, joined the West Midlands Fire Service in 1996 and Clive trained another black Lab, Odin, to be the first to join the LFB four years later. There are now twenty fire dogs working in the UK.

Sherlock (official title: Specialist Fire Investigation Dog) and his canine colleagues, Simba and Watson, have been trained to recognise 'target substances': ten different ignitable liquids including acetone and petrol that are used to start or accelerate blazes by arsonists. His skill is so refined that he is able to discern these substances even when they have evaporated or burnt in extreme heat (of up to 1,000 degrees), or been mixed with other liquids. He can even identify them in excess of a year after a fire took place.

During an investigation, Sherlock will pinpoint an area of interest using his nose, allowing human colleagues to document it and take samples for analysis to determine whether a fire has been started deliberately. His keen sense of smell is more accurate than any technology, and he is able to identify much smaller amounts of substance than any machine. This ability speeds up the process considerably, saving time and money both for the fire brigade and the police – for example, a room of ten by sixteen feet would take eight to ten hours to survey using a machine but can be covered by Sherlock in around two minutes.

Sherlock's handler is Station Commander Paul Osborne. The training both for the Cocker Spaniel and his handler is rigorous, as is the job. Fire dogs like Sherlock are selected when they are young 'based on enthusiasm and their drive for play'. Tenacious, long-nosed dogs such as Spaniels and Labradors are often the best fit. The dogs learn to work for rewards; for Sherlock, the ultimate prize is a tennis ball. Paul explains:

> We put a tiny bit of ignitable liquid on a bit of cardboard and then hide the tennis ball on top of that. So they will go into the room and look for their tennis ball, but at the same time they get the scent of something with an association with that ball.

Over a period of time, different scenarios are used and finally the ball is taken away when the dog knows exactly what is expected of him. After that, every time he identifies an ignitable liquid, a ball is his reward.

It can be perilous work. Although the dogs are never sent on to the scene until the fire has been extinguished and cooled, that doesn't mean that there aren't other hazards to contend with, so before a dog is deployed, the handler will always undertake a robust risk assessment to check for anything that could cause harm (like poisons or exposed electrical wiring) and reduce the risk. Fallen masonry, broken glass and other sharp objects are commonplace after a blaze, and Sherlock often has to wear special red boots to protect his paws. The three LFB fire dogs attend between 180 and 230 incidents each year, working not only in and around London, but supporting fire-and-rescue services nationally. In the history of dogs working in the brigade, no dog has come to harm, a testament to the handlers looking after the dogs' welfare.

Sherlock (also known as Mr Bustle Britches, Sherlockster Rockster or plain Rockster) lives with Paul, his wife and two children. Although he's a much-loved member of the household, he sleeps in a kennel outside and is only allowed in the kitchen. It's important that, as the pair are together all the time, there's a clear delineation between home and work.

His skill has been proven over and over. On one occasion when a rugby clubhouse in Surrey burned to the ground, the building was still burning when Paul and Sherlock arrived at the scene. Paul was able to recognise an unusual 'burn pattern' (which offers clues as to how a fire started and progressed). As the ruins cooled Paul identified an area for Sherlock to search. It was huge – around 3,000 square feet – but Paul had no doubt that the dog would get to the bottom of what had happened.

First Sherlock searched the perimeter and then, when the heat had died down, it was boots on and into the ruins. He quickly found traces of an accelerant on timbers at the rear of the building (later confirmed to be BBQ lighter gel) and, together with the burn pattern, this was enough to establish exactly what had happened. More than this, it was discovered that the same gel had been used in a number of other fires in pavilions and clubhouses across Surrey. Thanks to Sherlock, it became possible to link all the blazes to one suspect. Sherlock was the hero of the hour.

When we were doing the photo shoot for this book, Paul showed me Sherlock's powers of deduction. He placed a jar with paraffin in it (a year old so it had lost most of its aroma) in a bush in the garden. Sherlock set off at a rate of knots, tail wagging, examining every tree, flower, patch of grass and bush. He located the paraffin in under a minute and stood there, fixed to the spot, staring at it. All he wanted as a reward was his tennis ball. Paul says he is the most focussed, driven dog he has ever come across.

Sherlock's achievements – he has a 100-per-cent success rate – have seen him named London's most heroic working dog and earned him an Animal Hero award in 2017. Paul's pride in his dog is crystal clear: 'London Fire Brigade is here to help make London the safest global city, and really and truthfully, Sherlock does that every day he comes to work.' For Sherlock, it's elementary.

SHREK

We assume that sheep have a herd mentality because an entire flock can be bossed around by a sheepdog half their size, but the story of Shrek shows that the iron will of a sheep is not to be messed with.

Shrek was a merino born in 1994 on Bendigo, a sheep station on the South Island of New Zealand, where sheep outnumber people by almost ten to one. Smaller than sheep bred for meat, Merinos are kept for their fine, soft wool, which is renowned for its quality.

Merinos originated in southwestern Spain in the twelfth century, where they were instrumental in the country's economic development. So important was their contribution that, until the eighteenth century, exporting a merino from Spain was punishable by death. When the export rules loosened up, the merino sheep travelled to all parts of the world and careful cross-breeding programmes refined it to the modern merino we know today.

The humdrum life of a sheep on Bendigo was not enough for one young male. He didn't want to be one of 17,000. He wanted to be his own ram. Like something out of an Iron Maiden video, the sheep made his own 'run for the hills', heading for the high country of Central Otago. The rocky tors, remote plains and deep caves offered the perfect territory for a fugitive sheep to hide out and live off the land, though the extreme climate was not necessarily ideal. The region is the coldest and driest in New Zealand and in the freezing winters there is very little in the way of grazing. It was a mark of the sheep's steely disposition and extraordinary strength that he was able to survive for six years on the run.

When he was eventually found hiding in a rock cave in April 2004 by rangers out looking for other errant sheep, no one could quite believe their eyes. John Perriam, Shrek's owner, described him as 'looking like some biblical creature'. Dense wool covered more than three-quarters of his body – enough to make twenty large men's suits.

Shrek – named after the fictional ogre – was shorn live on national television, a process that took almost ten times longer than normal and yielded a fleece weighing a whopping sixty pounds – the average fleece is around ten pounds.

Shrek the fugitive sheep became a national celebrity, feted by the great and the good; he was even taken to parliament to meet the Prime Minister, Helen Clark. Over the years that followed, Shrek helped to raise over NZ $150,000 for children's medical charities.

SIMON

In 1949, British Royal Navy sloop HMS *Amethyst* was sent to keep order during the civil war on the Chinese mainland.

A rat infestation had been making life aboard the *Amethyst* both risky and unpleasant, so when seventeen-year-old

Ordinary Seaman George Hickinbottom saw the scrawny black-and-white tom cat at the docks of Stonecutters Island in Hong Kong, he offered him a job on the spot. He smuggled the stray aboard, keeping him hidden inside his jacket; and, later in his cabin, the hungry cat wasted no time in hunting down the rodents. He soon became a favourite with the rest of the crew and his rat-catching skills proved second to none. They called him Simon and taught him tricks, such as fishing ice cubes out of a jug of water.

The captain, Lieutenant Commander Bernard Skinner, was won over too and the pair formed a strong bond, often making their rounds of the ship together. Simon repaid his kindness with special 'gifts' (dead or bloody rats, dropped at his feet or occasionally left on his bed) and when the captain wasn't wearing his cap, the cat could usually be found curled up in it, sleeping peacefully.

As fighting between the ruling nationalist party of Chiang Kai-shek (the Kuomintang) and the Communist rebels of Mao Tse-tung's People's Liberation Army (PLA) intensified, the *Amethyst* was ordered to head from Shanghai to Nanking.

Mao's PLA held the Yangtze's north side, the nationalists the south. With a ceasefire between the two sides due to expire at midnight on 21 April, the British ship set sail immediately, although as neutrals in the conflict the crew were not expecting to be caught up in any trouble. But on 20 April, with around thirty-six hours of the ceasefire remaining and sixty miles still to go before they reached Nanking, the Communists opened fire without warning.

The *Amethyst*'s bridge, wheelhouse and engine room were rocked by explosions as the ship was hit by more than fifty missiles which killed nineteen of the crew including the captain and wounded twenty-seven more. Simon was nowhere to be seen. As the attack continued, a shell blasted a fifteen-foot hole in the bulkhead. Under imminent danger of sinking, the ship limped upstream to a small creek, where negotiations began for the release of the battered vessel and its crew.

No one saw Simon for several days, until he staggered onto the deck and was taken to the sick bay by Petty Officer George Griffiths. He was weak, dehydrated and in pain, and Medical Officer Michael Fearnley went straight to work removing shrapnel from his legs and back, stitching up his wounds and tending to the burns on his face. Simon's chances of survival were slim and the crew feared the worst.

Somehow he managed to pull through (although when his singed whiskers grew again they were badly bent) and soon got back to work, battling the huge and fierce rats that raided the food supplies and even attacked the crew. With the ship's boilers and fans shut down, the rodents had invaded the ventilation system and the situation was even worse than before.

A determined-looking Simon with sailors on board HMS Amethyst.

Simon's skill as a hunter remained important, but his work ethic also helped raise the morale of the men. They had been traumatised by the unprovoked attack, the loss of their comrades and the ordeal of having to bury so many friends and compatriots at sea.

While he made short work of most of the rodents, one remained undaunted. No trap could outsmart this huge and vicious rat, nicknamed Mao Tse-tung after the Communist leader. Simon finally cornered him in the storeroom, polished him off and proudly dropped the bloodied carcass by the men's boots as proof. The grateful crew promoted him to Able Seaman, and later he would also be awarded the Amethyst Ribbon for:

> distinguished and meritorious service . . . singlehandedly and unarmed [you did] stalk down and destroy 'Mao Tse-tung' [Mousey Tongue], a rat guilty of raiding food supplies which were critically short. Be it further known that from April 22 to August 4, you did rid HMS *Amethyst* of pestilence and vermin, with unremitting faithfulness.

Even with Simon's presence, life for the survivors aboard the *Amethyst* grew steadily more difficult. For almost three months they remained in limbo while the Communists demanded they admit that they had fired first and wrongly invaded Chinese waters. They began to run out of food and water, and rations were cut in half. There was very little fuel left and the situation soon became untenable. The new captain, Lieutenant Commander John Kerans, decided that the only way out was to make a run for it.

He seized his opportunity on the night of 30 July. There

was no moon and a passing merchant ship provided distraction for the *Amethyst* to begin the 104-mile journey back to the open sea. Fortune smiled upon them: five hours later they met the HMS *Concord* part way up the Huangpu River and were escorted the rest of the way to safety. After 101 days, the crew's ordeal was over, and the captain sent a message: 'Have re-joined the fleet south of Woo Sung. No damage or casualties. God save the King.'

As the *Amethyst* made its way back to Hong Kong, word spread of their ordeal and the crew, including Simon, were hailed as heroes. Their arrival in Hong Kong harbour was greeted by a media frenzy and, thanks to widespread coverage in newspapers and newsreels, the cat became an international celebrity overnight.

Before making the decision to award Simon a Dickin Medal, the PDSA got in touch with Captain Kerans. He sent a letter by return, which read:

For many days Simon felt very sorry for himself, nor could he be located. His whiskers, even now, show signs of the explosion. Rats, which began to breed rapidly in the damaged portions of the ship, presented a real menace to the health of the ship's company, but Simon nobly rose to the occasion.

Throughout the incident Simon's behaviour was of the highest order. One would not have expected a small cat to survive the blast from an explosion capable of making a hole over a foot in diameter in a steel plate.

Yet after a few days Simon was as friendly as ever. His presence on the ship was a decided factor in maintaining the high level of morale of the ship's company.

The medal was forthcoming, the only one ever to be awarded to a cat, and indeed to any animal serving with the Royal Navy.

Simon's heroics brought not only medals (he also earned one from the Blue Cross) but the adoration of the public. He received more than 200 pieces of fan mail a day as well as gifts of cat food and toys. At every port along the way, a welcoming committee would be out to greet him. He was placed in routine quarantine while plans were made for the medal to be presented by the founder of the Dickin Awards, Maria Dickin herself, as well as the Lord Mayor of London.

Sadly, Simon never made that date. His war wounds had left him with a weakened heart and he must have missed the shipmates he had done so much to help. Two weeks before his big day, he died at the tender age of two, leaving the crew and his many fans heartbroken.

Simon was buried with full military honours. The *Amethyst*'s entire remaining crew attended his funeral, and his gravestone, in the PDSA Animal Cemetery in Ilford, reads:

IN
MEMORY OF
"SIMON"
SERVED IN
H.M.S. AMETHYST
MAY 1948 – NOVEMBER 1949
AWARDED DICKIN MEDAL AUGUST 1949
DIED 28TH NOVEMBER 1949.
THROUGHOUT THE YANGTZE INCIDENT
HIS BEHAVIOUR WAS OF THE HIGHEST
ORDER

SIWASH

During the Second World War, one of the American forces' toughest campaigns was Operation Galvanic, fought in the central Pacific in 1943. The offensive took place midway between Papua New Guinea and Hawaii and involved 18,000 marines and one very determined duck.

You might ask what a common or garden waterfowl was doing in the middle of the Pacific Ocean. The answer is that earlier that year, in a bar in New Zealand, Sergeant Francis Fagan won a duck in a poker game and took her back to the United States Marine Corps (unofficially) to join up. Pet care was different then.

Siwash was named after his friend Sergeant Jack 'Siwash' Cornelius of Skagit County, Washington. Assumed to be a he, as so often happens, Siwash had to lay an egg to prove her true identity. She never left Fagan's side and before long became the 2nd Marine Division's (unofficial) mascot. Her commanding officer Colonel Presley M. Rixey told the *Chicago Tribune*, 'We value [her] too much to eat. Besides, we have no sliced oranges to serve with [her].'

The duck soon became known for her bravery. At the Battle of Tarawa on the island of Betio, when the marines attempted to storm the beach under a hail of bullets and bombs, Siwash followed, waddling into the fray and fighting with a Japanese chicken. After that, during the two fierce battles that followed at Saipan and Tinian, Siwash sagely kept watch from the boat.

There was talk of Siwash being awarded a Purple Heart,

but it never came to fruition. She did, however, receive the following citation (while still assumed to be a drake):

> For courageous action and wounds received on Tarawa, in the Gilbert Islands, November 1943. With utter disregard for his own personal safety, Siwash, upon reaching the beach, without hesitation engaged the enemy in fierce combat, namely, one rooster of Japanese ancestry, and though wounded on the head by repeated pecks, he soon routed the opposition. He refused medical aid until all wounded members of his section had been taken care of.

Siwash makes do with a bathtub in 1944.

When Sergeant Siwash returned to the States later that year, she received a hero's welcome. She helped to promote the sale of war bonds with Fagan and lived the life of a celebrity as a leading attraction at Chicago's Lincoln Park Zoo.

SMOKY

Little Smoky, weighing just four pounds and standing seven inches tall, was found in the New Guinea jungle by an American soldier in February 1944. No one knows how a Yorkshire Terrier (a breed specially developed to catch rats in the clothing mills of Yorkshire a century earlier) had ended up in the rainforest.

The campaign on the island was essential to the US Navy's drive across the Pacific and its liberation of the Philippines from the Japanese. The troops had to negotiate difficult and dangerous terrain, full of dense mangrove swamps, jungle and mountain ranges. There were no roads or railways and when they tried to push through the jungle in jeeps, there were constant mechanical problems. It was one of these that led to the discovery of Smoky. While soldiers were working on the engine of a stalled jeep, they heard a noise coming from the ground and discovered it was the tiny dog, hiding in a foxhole.

A soldier took her back to camp where one of his comrades shaved off her silky coat, thinking she'd be less hot without it. Sharing the tent was Corporal Bill Wynne of Cleveland, Ohio, who agreed to buy the dog for the steep price of AUD $2 (a significant percentage of his pay). She was worth every cent as this was the start of a long and happy friendship that would change both their lives. He called her Smoky.

Tiny Smoky fits comfortably into an American soldier's helmet.

The little dog spent much of the next two years tucked in Wynne's backpack as he trekked through the jungle and took part in combat fights in the Pacific. Smoky slept in Wynne's tent, on a blanket made from the baize from a card table. She shared his C-rations – the tinned wet food given to troops when fresh or other packaged food was not available – and the odd tin of Spam. Together they faced the ever-present danger of snakes – which would have made short work of her – and frequent air raids, more than 150 in total (plus later a typhoon in Okinawa).

Somehow, Smoky not only survived but also remained in pretty good shape – even after descending thirty feet from a tree in a parachute specially made for her. At a time when many of the American troops in the southwest Pacific were suffering from psychological trauma, she offered Wynne a much-needed distraction from homesickness and stress.

When Wynne was hospitalised with dengue fever, Smoky soon charmed the nurses into letting her stay in his room.

They realised that she could cheer up the other injured and sick patients so took her on their rounds during the day before letting her sleep on Wynne's bed at night. In a way, she was the first military therapy dog.

Away from the front lines, Wynne taught Smoky how to dance the jitterbug, 'sing', spell out her name, walk a tightrope, ride a scooter and play dead on command. Everywhere she went, Smoky charmed the troops. And on one occasion, she would save Wynne's life.

Bound for Luzon during the invasion of the Philippines, Wynne and Smoky stood on the landing deck of their transport ship and watched a kamikaze attack unfold before their eyes. Explosions roared. A nearby ship was hit and anti-aircraft fire boomed all around them. The terrified dog got her owner to duck, as a shell flew over their heads, wounding the eight men standing beside them. Both Smoky and Wynne escaped unhurt. Wynne described Smoky as 'an angel from a foxhole'.

Alongside the rest of her comrades, the dog earned eight service or 'battle' stars for being part of conflicts involving grave danger or death. Her most famous feat was when she helped engineers rebuild a critically important airfield in the Lingayen Gulf on Luzon. Constant bombardment from the Japanese had battered the communication systems and, to mend them, the Signal Corps needed to run telegraph wires through a seventy-foot-long pipe just eight inches in diameter. It was essential work – the base served three separate squadrons – but it would take engineers days of digging to repair it by hand. Soil had seeped in through the joins, meaning there was only a four-inch space in some places. Luckily, they had a tiny terrier on their side.

Wynne tied string to the wire, then attached it to Smoky's

collar. He ran to the far end of the tunnel and called her. It was dark and narrow and after only a few steps, she turned back. Wynne encouraged her to try again, but when the string snagged, it looked as though the task would be impossible:

> By now the dust was rising from the shuffle of her paws as she crawled through the dirt and mold and I could no longer see her. I called and pleaded, not knowing for certain whether she was coming or not. At last, about 20 feet away, I saw two little amber eyes and heard a faint whimpering sound . . . at 15 feet away, she broke into a run. We were so happy at Smoky's success that we patted and praised her for a full five minutes.

The tiny dog was thus credited with saving forty planes and the lives of 250 men.

At the end of the campaign, having been smuggled into the USA in a modified oxygen mask, Smoky became a media sensation. She visited veterans' hospitals across America and travelled all over the world to demonstrate her skills; she was a regular on TV specials, performing tricks on more than forty-five live shows.

When she died in 1957 at the age of fourteen, she was buried in Lakewood, Ohio, where today a statue of the dog wearing a GI helmet stands in memory of 'Smoky, the Yorkie Doodle Dandy, and the Dogs of All Wars'.

STOFFEL

Honey badgers, or ratels, are widely found in Africa, southwest Asia and on the Indian subcontinent. They have a black underbelly and lower half of the face with white on top, a bit like a skunk, to whom they are related along with weasels, otters, ferrets and other badgers. What sets honey badgers apart are their thick skins, their strength, their defensive abilities and their intelligence. They have few natural predators and they are masters of escape. This is an animal that can get itself out of any situation.

The South African military has a vehicle named after them and, in the 2002 edition of the *Guinness Book of World Records*, the honey badger is listed as 'the world's most fearless animal' – a description that could not be more apt when it comes to our hero, Stoffel.

Stoffel began life in domesticity, reared by a farmer, but after he created chaos in the farmer's house, he was taken to the Moholoholo Wildlife Rehabilitation Centre near the Kruger National Park in South Africa. When he first arrived, the badger was allowed to roam free with two other adult (female) honey badgers but he did not behave well. He went on a murder spree, killing rabbits, small bucks, even a tawny eagle. He was aggressive to humans, chasing staff out of the kitchen at the lodge so that he could help himself to all the food he fancied as they fled. He was a handbag-snatcher, tearing open bags in the cloakrooms to see what lay inside.

While the better-behaved females were allowed to return to the wild, Stoffel was considered ill-equipped for life on the outside because, as well as being hand-reared, he had an impaired sense of smell and his ability to find his own food was limited. The trouble was that he wasn't really adjusting to life on the inside either and he kept managing to escape from his camp.

Staff at the centre tried everything they could think of to keep him inside. No one expected Stoffel to be able to undo the locks that fastened the gate, but he did so thanks to a girlfriend acting as his accomplice: she climbed on top of him to undo the top latch while he undid the bottom one. Next, wire was used to secure the locked gate, but Stoffel quickly set to work unwinding it and opened the latches. Out he went again.

To protect the other animals, staff at the centre put him in a quarter-hectare camp filled with grass and trees. When he broke out to fight the neighbouring lions (fifteen times his size), he ended up in hospital for the next two months; he could have been killed. It seems he is not one to learn from his mistakes: the minute he got out, Stoffel tried to get back to the lions, no doubt to sort 'em out.

The local rotary club sponsored a new enclosure built from brick. The walls were too high and smooth for Stoffel to scale so surely this would keep the breakout badgers in their place? Not a bit of it. Stoffel dug under the walls. When measures were taken to stop his digging, he climbed up a tree, bending its branches to reach the wall, and simply walked out. When a keeper left a rake in his enclosure, he moved it to the wall and climbed up it. He also tried rolling rocks to the wall, piling them up with his strong hind legs, and then using them to leapfrog out. He even made himself an escape route by means of a mud ramp.

Stoffel is so ingenious that Brian Jones, the founder of the

Rehabilitation Centre, is in awe of him. He told the BBC: 'The intelligence is just beyond anything . . . Every time I devised some plan it was like a game for him to work out how he could get over this.' People wondered if Brian had trained him to escape. Brian exclaims, 'Train him? . . . Not at all. I didn't even think of it. He outwitted us each time with his schemes.' With Stoffel's great escapes caught on video, he went viral and is now a YouTube sensation. If you don't believe me, look him up and you'll be one of 30 million or so who have seen his Houdini talents.

Now, the badly behaved badger is an ambassador for his species, attracting visitors from far and wide. Money raised from ticket sales is ploughed back into the centre's essential work – and its tuck shop has been renamed in his honour. While this Harry Houdini of the badger world has undoubtedly caused untold headaches for his keepers, Stoffel has also helped raise essential funds for wildlife rehabilitation.

STUBBY

When a dog with a barrel shape and short stature wandered into the Yale University football stadium in New Haven, Connecticut, few would have predicted that such an unprepossessing stray would go on to become the most decorated dog in American history.

It was 1917 and the 102nd Infantry Regiment were drilling, in preparation for going to war. The dog was spotted by a private named J. Robert Conroy, who took a liking to the puppy and decided to keep him. He named him Stubby on account of his short legs and tail, and they soon became inseparable.

Variously described as a Boston Terrier, an American Bull Terrier or a dog 'of uncertain breed', Stubby was very intelligent. He quickly learnt the bugle calls and drills and would even put his paw to his eyebrow in imitation of a salute.

As the unit left for France, Conroy couldn't bear to be separated from him so he smuggled Stubby aboard the troop ship SS *Minnesota* in his greatcoat, then rushed him down to the hold where he hid him in a coal bin. The commanding officer was not best pleased to find a small dog in their midst, but he was charmed into submission when the dog saluted him.

Stubby joined the men on the front lines in February 1918, where he quickly proved his worth. On one occasion he heard the whine of an incoming shell before any of his comrades and was able to warn them to take cover just in time. He could locate survivors who had been wounded on the battlefield and help lead the medical team to them. Dogs have fifty times more olfactory receptors in their noses than humans and Stubby's quick thinking and bravery meant that when he smelt lethal fumes, he ran up and down the trenches barking loudly and nipping at the soldiers until they woke up. His keen nose saved many of his fellow soldiers from asphyxiation as they slept. For this life-saving action, he received his first military ranking: private, first class.

Gas attacks were one of the deadliest hazards of the First World War. Mustard gas became the most widely used and could kill by blistering the lungs and throat. Stubby had already been injured by the gas and was hyper-alert to any future danger. He was given his own specially designed gas mask, although the *New York Times* reported that 'Stubby's physiognomy was of such peculiar contour that no mask could afford real satisfaction.'

Stubby's presence on the battlefields of Europe was an unusual one. Although more than 50,000 dogs served during the course of the war, it was not a common policy for American troops to use them on the actual battlefield. At one point, the US Army borrowed a number of French military dogs but, as the dogs were unable to respond to commands in English, it didn't work out.

The 26th (Yankee) Division saw more fighting than any other American infantry division and Stubby was there throughout, with the full support of regimental leader Colonel John Henry Parker. It was said that the dog was the only member of his regiment who could talk back at him and get away with it. According to the Associated Press, the dog's rage was 'so savage' that 'it was found necessary to tie him up when batches of prisoners were being brought back, for fear that trouserless Germans would be reaching the prison pens'.

His first month in combat, beginning at Chemin des Dames, north of Soissons, was spent under constant fire and then Stubby was wounded in the foreleg by a German grenade. As soon as he recovered, he returned to his unit and became adept at finding wounded men in no-man's-land. He would bark to alert the medical team, but if the men were beyond help, he would stay with them to comfort them as they died.

Stubby also seemed to have worked out who was friend or foe. One night, investigating a sound, he crept out of the trenches to find an enemy spy attempting to map the layout of Allied trenches. The *New York Times* reported that 'attempts by the German to deceive the dog were futile. Seizing the prisoner by the breeches, Stubby held on until help arrived.' The commander was so impressed that he nominated Stubby for promotion, while the men confiscated the German's Iron Cross and awarded it to the dog.

In all, Stubby spent 210 days on the battlefield serving in four offensives and seventeen battles on the Western Front, where he suffered further injuries to the leg and chest.

After the Battle of Château-Thierry in July 1918, when he managed to avert yet another deadly gas attack, the women of the liberated town showed their gratitude by making him a chamois coat (on which his many medals would later be pinned). He became the only dog to be nominated for rank and then promoted to sergeant through combat.

Sergeant Stubby, in his decorated chamois coat, visits the White House in 1924.

Despite all this, the official ban on soldiers' pets meant Conroy still had to smuggle him home. Once back in

America, Stubby received the hero's welcome he was due and was made a lifetime member of the American Legion and the YMCA (the latter entitled him to 'three bones a day and a place to sleep' for life).

Conroy went on to study law at Georgetown University and of course the dog went too. In fact, he was adopted as mascot.

Stubby met US presidents Woodrow Wilson (to whom he offered his paw), Calvin Coolidge and Warren G. Henning and, in 1921, five years before he died, was awarded a gold 'Hero Dog' medal by the Humane Education Society at a ceremony in the White House. General John Pershing, commander of the American forces in Europe, spoke solemnly of Stubby's 'heroism of highest calibre' and 'bravery under fire'. The *New York Times* reported that Stubby 'licked his chops and wagged his diminutive tail' in response. Other awards include a New Haven WWI Veterans Medal, the Purple Heart, the Republic of France Grande Guerre Medal, the St Mihiel Campaign Medal and three service stripes. His obituary in the *New York Times* was twice the normal length, running under the headline: 'Stubby of A.E.F. Enters Valhalla'.

TACOMA

What's not to love about dolphins? Their smile makes them look permanently happy, their athleticism is a joy and they are hugely intelligent. They are also extremely sociable. They have a very mature and responsible sense of parenting, keeping their young close and teaching them skills that will help them live as independent adults. If they were shown how to drive a car, cook and do their own laundry, I reckon they could.

Along with humans, apes and elephants, the brains of dolphins contain spindle or von Economo neurons, which are associated with traits such as emotion, communication, perception and problem-solving.

The dolphin can also do one thing we can't – use its hearing to see. Light does not penetrate the very depths of the ocean but sound, which travels more easily, does. So dolphins use echolocation – literally using sound to find things – which allows them to 'see' in conditions where there is little or no visibility. Sounds, or clicks, produced by the dolphins are sensed as echoes in their jawbones and the information is then transmitted to the brain. The length of time that an echo takes to bounce off an object can establish not only how far away it is, but also its size and shape. The process is so efficient that it allows dolphins – which can dive much deeper than humans – to find things buried in up to half a metre of mud on the seabed. With skills such as these, which can better even the most sophisticated pieces of machinery, it's not surprising that dolphins have been co-opted by the naval superpowers of the world.

The Americans began their Marine Mammal Program in 1960, studying not only dolphins' sonar capabilities that might be used for locating anything from underwater mines to lost equipment, but also the aerodynamics that allow them to swim at such great speeds and depths. They also trained them to hold cameras in their mouths, allowing them to conduct underwater surveillance operations. The project peaked at the height of the Cold War when millions of dollars were ploughed into research and training to try to top what the Soviets were doing with their dolphins. The Spy Who Came in from the Cold became the Spy Who Came in from the Sea.

Between 1965 and 1975, five US dolphins were used to guard boats during the Vietnam War, although no one was allowed to tell you that – their presence was classified information. Dolphin patrols searched for enemy divers sent to plant explosives and, get this, could even place restraining clamps on intruders.

Training a dolphin takes about a year and is a major investment. By the end of their education programme, each one is worth around $2 million and can serve their country for up to twenty years. We are, of course, dealing with a wild animal so it doesn't always work out, and sometimes that $2-million investment will make a bid for freedom at its first opportunity in open water. There is also the risk that the territorial instincts of a dolphin will take over and the indigenous pod will drive away the navy spy dolphin from its territory.

Dolphins live for between thirty and fifty years so a twenty-two-year-old is at the peak of his powers, and that's the age Tacoma was when he was flown into Iraq's only deep-water port in 2003 to help clear the area of explosives. The US Navy Explosive Ordnance Disposal Mobile Unit Three (Eodmu 3) was tasked with making Umm Qasr Port safe for ships carrying vital humanitarian aid.

Tacoma was trained in a massive tank on a US warship and then transported by helicopter using a special travelling sleeve. It was the first time he had been used to clear mines. His handler described him as 'vocal' but 'one of the best at his job'.

The waters in that part of the Persian Gulf, just north of

Kuwait, offer almost zero visibility, making it impossible for human divers to search for explosives. That's where dolphins' echolocation skills come to the fore.

A Bottlenose dolphin, wearing a locator, leaps out of the water during a training session in the Persian Gulf.

The mines had been in the harbour since the Gulf War in 1991 and many had sunk deep into the silt, but they were still live and could be triggered by the approach of any vessel. They don't even need to be touched to set off an explosion – it can happen because of shifts in the magnetic field around them caused by the steel hull of a ship or the alloyed steel of a submarine.

As dolphins' bodies contain no metallic elements, Tacoma was able to get close to the mines without triggering them. The Atlantic Bottlenose dolphin went right to work. Now for the really clever bit. He had been trained not to touch the mines, so if he found one he would swim up and press a plastic ball on the front of the boat. Special buoys would mark where it was and then the navy divers could swim down and neutralise the mine.

The mission was not without its panic points and on 4 April Tacoma went missing for forty-eight hours. No one knew whether he had been tempted by the open seas, driven off by Iraqi dolphins or killed in action. Mercifully, he returned unharmed to continue his life-saving mission. Along with four other dolphins, Tacoma located more than a hundred underwater devices. It was thanks to the efforts of the dolphin detection team that the British aid ship *Sir Galahad* was able to sail safely into port to deliver essential supplies for many thousands of Iraqis. Commodore Brian May of the Royal Navy said, 'The Lord God decided to give the dolphin the best sonar ability ever devised. We can only aspire to their ability.'

TAMA

Cats have a special place in Japanese culture. They are believed to bring good luck and they symbolise prosperity and fortune. Whether in ancient art or the modern phenomenon of Hello Kitty, cats have consistently been centre stage. In Japan, there are at least two islands known as 'cat island'; there is a 'cat town' within Tokyo and there is such a thing as a cat café where you can pet cats while drinking your coffee or tea, take a cat souvenir home if you want to, and, well, you can imagine the endless possibilities of cat-related tat.

When I was choosing my top feline heroine, Tama stood out. The belief that cats can bring good fortune really was true in her case as she is credited with saving the local economy in Wakayama in western Japan. She did so by holding down the unlikely job (for a cat) of stationmaster at Kishi Station.

The calico kitten was born in Kinokawa in 1999, one of a group of strays who would hang out close to the station. She was informally adopted by the owner of a local convenience store but still liked to spend time near the trains. Her friendly nature quickly made her a favourite with commuters who began to call her 'stationmaster' Tama.

The line served by Kishi Station into Wakayama city was under financial pressure and was earmarked for closure in 2004. The locals managed to keep it going but, two years later, all staff members were removed to cut costs. So there were no ticket collectors, no guards and no stationmaster. That's when Tama stepped up to the plate. She moved into her own office (a converted ticket booth, complete with litter tray) and became the face of the railway. Her duties included greeting commuters from a table by the ticket gates. Her 'salary' was all the cat food she could eat, plus a gold tag for her collar, engraved with her name and position. She was also given a summer hat to wear in warmer weather.

She was adored by commuters and railway staff alike. Even the president of the Wakayama Electric Railway Company fell for her charms and ordered her a customised stationmaster's hat. In January 2007, he officially pronounced her the first feline stationmaster in Japan.

Tama appeared in the station's promotional material and became a media star. In 2007 she won the railway's Top Station Runner award, and her year-end bonus was a cat toy and some morsels of crab. In 2008 she was promoted to 'super station manager', making her the company's only female in a managerial position. Her name tag was modified to include an 'S' for super and her portrait was commissioned.

Tama was later knighted by the prefecture governor 'for her work in promoting tourism' and given a dark blue ceremonial gown with white lace ruffles at the neck. Tama-mania swept the nation as thousands of tourists flocked to Kishi to witness the stationmaster at work. A study showed that in 2007 alone, the cat's presence attracted 55,000 additional passengers on to the Kishigawa Line.

Tama's star continued to rise and she was promoted to operating officer in recognition of her part in the expansion of the company's customer base. This made her the first cat to become an executive of a railway company. Her mother, Miiko, and sister, Chibi, were taken on as her assistants. The station building and one of the trains were remodelled by Eiji Mitooka, the man behind the bullet trains. The designer train has carriages decorated with paw prints and cartoon images of Tama. At every stop, a recording of meowing is played as the doors open. The front of the train sports whiskers.

Tama went on to become managing executive officer, making her the railway's third most important employee after the company president and managing director. A year later she was made honorary president of Wakayama Electric Railway Company for life.

She died of heart failure at the age of sixteen, in 2015, by which time her contribution to the local economy was more than JP¥1.1 billion (or £7.8 million). An estimated 3,000 people attended the funeral of the 'Honourable Eternal Station Master' at Kishi Station, bearing bouquets and tins of tuna. Tama was elevated to the status of Shinto goddess.

The UK also had its own high-ranking feline public servant in the giant shape of Tibs the Great, who was officially employed at Post Office headquarters to keep rodents under control, and therefore keep documents and mail safe from damage and destruction. The role carried great responsibility and a wage of 2s. 6d. a week. The level of Tibs's pay was discussed in Parliament in 1952 following concern that it hadn't risen in line with inflation.

The Assistant Postmaster General, David Gammans, responded to claims of unfair pay:

> There is, I am afraid, a certain amount of industrial chaos in the Post Office cat world . . . It has proved impossible to organise any scheme for payment by results or output bonus. These servants of the State are, however, frequently unreliable, capricious in their duties and liable to prolonged absenteeism.

He added that 'there has been a general wage freeze since July 1918, but there have been no complaints', and then pointed out that all Post Office cats were offered equal pay and given 'an adequate maternity service'.

Tibs was never unreliable and served the Post Office without fail for fourteen years. When he died in 1964, a number of newspapers ran obituaries. He weighed a whopping twenty-three pounds, which may not have been entirely due to his rodent-catching prowess. He had, in later life, decided to move his office to the staff dining-room. Now that's a clever cat.

THE TAMWORTH TWO

The shadow of death hung over two Tamworth pigs. At the tender age of just five months, their end was nigh, their destiny sealed – or so it seemed. However, these two brave piglets had other ideas.

Their slaughter was marked for 8 January 1998. The day dawned with the young brother and sister being loaded into a lorry heading for the local abattoir in Malmesbury, Wiltshire, when they decided to make their great escape. They ran away from the abattoir, squeezed through a fence and swam across the River Avon before dashing through nearby gardens into a dense thicket near Tetbury Hill. That's where they hid out for the following six days.

The tale of a pair of piglets avoiding the butcher and emulating the adventures of Butch Cassidy and the Sundance Kid was lapped up by the media. They were dubbed the 'Tamworth Two'. Reporters from all over the world, including Japan and America, were sent to cover the story.

The plot thickened when the pair's owner, council road sweeper Arnoldo Dijulio, declared that once the pigs were recaptured, he'd be sending them straight back to the slaughterhouse. Animal lovers and newspapers alike were quick to offer large sums of money to save the pigs' bacon. The *Daily Mail* eventually secured a deal to save 'Butch Cassidy and the Sundance Pig' in return for exclusive rights to their story and the photographs.

After a week of freedom, the pigs were spotted foraging in the garden of Harold and Mary Clarke. Butch was recaptured but Sundance gave everyone the slip once again and headed back into the depths of the thicket. It took two spaniels and a couple of tranquilliser darts to catch him. The vet who checked him out described him as none the worse after his big adventure, but took the precaution of stabling him behind a six-foot wall and a few bolts, adding, 'He's obviously quite bright. He's foxed a number of people for a number of days. I don't want to spend another day chasing around Malmesbury.'

The *Daily Mail* then sent Butch and Sundance to live at the Rare Breeds Centre, an animal sanctuary in Kent, where many of their fans were able to visit them. The BBC even made a drama about them in 2004 called *The Legend of the Tamworth Two.*

Butch and Sundance died within six months of each other in 2011 at the age of fourteen.

THANDI

I was filming at the Kariega Game Reserve in South Africa for a BBC programme called *Operation Wild* when I met Thandi in 2013. Her name means 'courage' and 'to be loved' in the local Xhosa language. She was shy and hiding behind a bush, but we knew it was her because she didn't have a horn. A year earlier, Thandi had been one of three rhinos brutally mutilated by poachers who had sedated them with darts and removed their horns with a machete, leaving them in pools of blood.

'Their situation broke my heart', Dr Will Fowlds said when

he arrived on the scene, admitting, 'I just sat in my vehicle and wept for them and all the other rhinos out there who were being mutilated on a daily basis.' One rhino was already dead and another died a few months later. Fowlds continued to treat Thandi. 'I had already witnessed something of the drive in her to stay alive', he says. She needed multiple skin-graft operations, some of which proved hugely problematic. Her plight touched millions of people across Africa and around the world and, as a result, she has become the figurehead of the fight against poachers.

The desire for rhino horn is driven by the fallacy that somehow it can provide a remedy for everything from fever to gout. It is wrongly believed to work as an aphrodisiac and a hangover cure. This false belief has led to rhino horn being hugely valuable. It is genuinely worth more than its weight in gold – around double the value. Rhino horn is made of keratin, the same substance found in human hair and nails, so you'd be just as effectively 'treated' with powdered fingernails and hair tea.

Over the last century, poaching has decimated the rhino population in Africa and Asia. From half a million in the wild, there are now fewer than 30,000 and three of the five species are critically endangered. In South Africa, the Western Black and Northern White rhino are extinct in the wild. Poaching is a particular problem in South Africa, where almost 80 per cent of the world's remaining rhinos can be found and where one is killed on average every eight hours. It is also a hugely dangerous environment for keepers and rangers, over 900 of whom have been killed since 2009 while protecting wildlife.

The film I was making followed a process of safely injecting a rhino's horn with a pink dye that would permanently colour it and render it poisonous to humans. It's similar to the system used on banknotes and the dye shows up in airport scanners, even if the rhino horn is ground into a powder. Signs were put up around the game reserve in various languages, including Mandarin, so that poachers and their overlords might understand that the rhino horn was not worth taking.

I never stopped thinking about Thandi and the image of her lying in a pool of blood, her horn hacked off. Amazingly, she not only survived but also, not long after I had seen her, blood tests revealed that she was pregnant. Fowlds cried again when he heard the news. It felt like a miracle, confirming to him that she was a beacon of hope. 'This rhino has changed my life', he said:

> I can't say it's for the better as I could never wish to fight a war such as this one, but she has shown me inner strength which I must follow. She has inspired action in myself and many around me which I must continue. She now celebrates life and with it the hope that against all odds, we can and we WILL overcome the massive challenges that threaten to take them down.

Nearly three years after the brutal attack, Thandi gave birth to a healthy female named Thembi (or 'hope'). She has since gone on to have two more calves: a boy named Colin in honour of Colin Rushmere who founded the Kariega Reserve and, in 2019, another boy named Mthetho which means 'justice'. This was particularly appropriate as he was born in the same week that members of a notorious poaching gang (who may well have been responsible for the attack on Thandi) were each given twenty-five-year prison sentences.

Thandi was the first rhino ever to survive a poaching attack, and her fight for survival and determination to bring new life out of her dark experiences inspired people around the world. It made a massive impact on one of the reserve's volunteers, Angie Goody, a beef and sheep farmer from the Isle of Man. She made a promise to devote herself to the fight against poaching and since then has been fully committed to 'conservation, raising awareness and fundraising for the rhinos of South Africa'. She set up Thandi's Endangered Species Association and has raised many thousands of pounds to help fund the rhino's ongoing treatment, as well as buying equipment to help in the fight against poachers.

Simon Jones, CEO of Helping Rhinos, an international charity working to help the species survive at sustainable levels in their natural habitat, revealed, 'Thandi has played a significant role in my life and is a huge part of why Helping Rhinos exists today.'

I was deeply affected by Thandi and have since heard Dr Fowlds speak at the Royal Geographical Society. He is a passionate and engaging orator as well as a brilliant vet, but he will always maintain that his inspiration comes from the example of a remarkable rhino whose bravery has done so much to raise awareness of the species' magnificence:

> Thandi continues to show us the value of life and a species which deserves our hearts. [Her] story of sheer determination and will to survive represents hope in the face of hopelessness. This story stands as testimony to the worst and the best of human attitudes towards animals.

THULA

There is plenty of research to suggest that having a pet can act as a calming influence on children with neural diversity. It doesn't work for everyone but when it does, the impact is extraordinary. I have seen dogs, horses and guinea pigs offer comfort and instil a sense of responsibility as well as giving unconditional love to children who struggle with boundaries and emotions. It is rare for a cat to fall into this category because (and I say this as a cat 'owner' myself) they tend to do things on their own terms. You don't ever master a cat. Instead, you are their servant.

There is, occasionally, an exception to that rule and Thula is a special cat who has had a transformative effect on a little girl called Iris.

Iris was different from other toddlers. She didn't like to make eye contact or engage with anyone around her. She would become distressed if she was near people she didn't know. Her mother, Arabella Carter Johnson, was concerned so in 2012 she took her to see specialists. Iris was diagnosed as severely autistic and the doctors warned she might never talk. 'Reading everything I could find on autism, I soon realised there were no quick fixes', Arabella said. She describes how 'almost everything could provoke a meltdown – the clatter of the dishwasher or someone waving a toy in her face. Unable to cope with the noisy, complicated world, Iris began to shut it out – and she became almost totally silent.'

Arabella tried getting a dog as a way of helping Iris

communicate but explains that 'Iris and the dog didn't get along – Iris hated being licked and the tail wagging, the hyperactivity of the dog would upset her.'

By chance, a cat came into their lives when her brother went away over Christmas and asked her to catsit. Her parents were worried about how Iris would react but she and the cat seemed to bond. Arabella decided that it was worth getting a cat of their own and her investigation into different breeds led her to the Maine Coon.

Maine Coons are the largest breed of domestic cat and can weigh up to eighteen pounds. There are all kinds of stories about where they come from. Some believe that they are wild cats crossed with racoons, others that they are descended from six cats that Marie Antoinette loaded onto Captain Samuel Clough's ship as she tried to escape revolutionary France in 1792. Unlike their unlucky royal owner, the cats crossed the Atlantic to land safely in Maine, where they became official state cat in 1985. Their sociable, loyal and intelligent nature has led to them being described as 'the dogs of the cat world'.

Thula the Maine Coon arrived at the Carter Johnson house and spent her first night sleeping in Iris's arms, unperturbed when the little girl constantly stroked her whiskers and tail. 'Thula loved all the things that Iris found difficult', said her mother. One of those things was the sensation of anything – including clothing and water – touching her skin. This had made bath time a nightmare for her parents, until Thula came along. Unlike other cats, Maine Coons love water so Thula hopped in too. 'It was like heaven', said Arabella.

Thula was a typically mischievous kitten, but around Iris, she instinctively altered her behaviour. If Iris was stressed or impatient, the kitten would sit on her lap to calm her down.

If she woke in the night, Thula would comfort her and stay until she had settled. Iris began to speak to her pet, asking her to sit or to follow her round the house. In turn, the kitten kept Iris calm and happy during activities from picnics to shopping, even mimicking Iris's actions as she played and painted.

With Thula helping Iris's confidence and concentration, her painting, initially introduced to help with speech and behavioural therapy, soon took on a life of its own. Iris loves colours and has a gift for combining them with beauty and flair. After her mother started posting her paintings on Facebook, they became something of a sensation and one was even bought by the actress and activist Angelina Jolie. The proceeds are put towards Iris's education and therapies. Arabella says:

> I never imagined an animal would bring so much joy to Iris's life . . . Before Thula, she used to cry a lot – it was distressing for both of us – but she has changed more than I could have imagined since Thula came into our lives. It is a miracle.

So here's to Thula and all the animals like her who make miracles happen in everyday life.

TIRPITZ

A pig that survived a sinking ship, jumped sides in the First World War to join the British Navy and ended up raising thousands of pounds for the Red Cross is not something you come across every day, but then this is a book about exceptional animals.

Tirpitz was resident on SMS *Dresden*, one of many German warships that sailed into battle with a pig on board. She was not there as a mascot or a stand-in admiral but for the practical Teutonic purpose of offering fresh meat when supplies ran low.

The *Dresden* had been ordered into the South Atlantic to join up with the forces of Vice Admiral Maximilian von Spee. After one successful battle, the Germans were defeated at the Battle of the Falkland Islands. Although the *Dresden* was the only ship that got away, she was eventually tracked down in Cumberland Bay, near what's now known as Robinson Crusoe Island.

The captain wasn't able to find a safe harbour away from the guns of the British fleet, so he decided to scuttle his vessel. The crew rowed ashore in lifeboats, leaving most of their equipment and possessions behind. Tirpitz the pig was the only living creature left aboard the *Dresden* as it sank.

The sow's survival instinct helped her find her way above deck and jump into the sea to swim to safety. She was spotted, about an hour later, by a petty officer aboard HMS *Glasgow*. When he jumped in to save her, the panicking pig almost drowned him. However, with the help of a winch, both were eventually hauled on board, amid laughter from the crew. The pig seemed happy to swap sides to join the Allies. The crew adopted her and made her their mascot, naming her Tirpitz after Alfred von Tirpitz, the secretary of state of the Imperial Naval Office and admiral of the German navy. As an added irony, they awarded her the Iron Cross, the top German military medal for bravery.

Tirpitz relaxes on board HMS Glasgow.

Tirpitz had found her family and they weren't planning to eat her. She remained aboard the *Glasgow* and, after quarantine, was safely installed with other farmyard animals at the naval training centre on Whale Island in Portsmouth. Tirpitz was a large sow and she put her size to good use when she destroyed the chicken run in search of food. She was becoming a problem and was moved along to the former commander of HMS *Glasgow*, Captain John Luce.

Captain Luce wasn't overwhelmed by their reunion and offered Tirpitz to a charity auction raising money for the British Red Cross. So, the pig who had survived a shipwreck and seen war action was bought for 400 guineas (around £20,000 in today's money).

Tirpitz may well have been sold again; she eventually ended up in the ownership of the 6th Duke of Portland. She died in 1919 and the duke had her head mounted. In 1920 he presented it to the newly opened Imperial War Museum in

London. You can still see Tirpitz there today in the First World War galleries. Her trotters were made into the handles of a stainless-steel carving set that was presented to the next HMS *Glasgow*.

Another brave hog, known only as 'Pig 311', hit the headlines in 1946 after the United States conducted nuclear weapons tests in Bikini Atoll, a reef in the Marshall Islands in the North Pacific.

Operation Crossroads was devised to test how nuclear explosions would affect ships at sea and involved twenty-two boats, all moored at different distances from the blast. Aboard them were animals including guinea pigs, mice, pigs, goats and rats – all sacrificed in order to understand the effects of nuclear fallout.

The force of the blast destroyed a number of ships and, with them, a third of the animals; many more died of radiation sickness over the course of the next few weeks. However, a lone pig, found swimming in a lagoon by recovery crews, somehow survived. The super-surviving swine was studied closely by the navy for a further three years along with Goat 315, the only other animal to make it through the operation unscathed.

Later the pair were given to the Smithsonian National Zoo. Attempts to breed Pig 311 came to nothing – many believed that the radiation had left her sterile.

TOGO

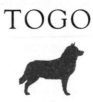

The city of Nome was once Alaska's largest when the gold rush brought thousands of prospectors. Nowadays, there are just under 4,000 inhabitants. In winter, the temperature can dip below –40°F and can leave the population cut off from the outside world.

That's what happened in January 1925 when the city was ravaged by a diphtheria epidemic. Nome had just one doctor, Curtis Welch, along with four nurses and a hospital with a mere four beds. They had run out of up-to-date supplies of the required antidote and the replacement order had not arrived. By early January four children had died.

Diphtheria is a deadly bacterial infection affecting the mucous membranes of the nose and throat. It can strike anyone, but young children and adults over the age of sixty are particularly at risk. It can lead to breathing difficulties, paralysis, heart failure and death. In 1921, there were more than 200,000 reported cases in the United States, and over 15,000 of these patients died.

Nome was put in quarantine but as the numbers grew it became clear that the situation was very grave. With 10,000 people in the local area at risk and a predicted mortality rate that could wipe out most of them, Welch grew desperate. He sent a telegram to the US Public Health Service in Washington begging for help, but heavy snow had left the roads impassable and ice prevented any ships from getting close. The freezing temperatures meant the only planes operating in the area – three vintage biplanes with open cockpits and water-cooled engines – could not fly. Nome was completely cut off. The only answer was to turn to dogs for help.

The dogs in question were Siberian Huskies, a breed originally from northeast Asia who were used to freezing conditions. They had been introduced to Nome during the gold rush years as sleigh dogs and were fast, strong and hardy.

Fourteen years after they had helped Roald Amundsen become the first man to reach the South Pole, Siberian huskies were again the secret to success. Togo was the lead dog in the sleigh pack that set out from Nome to meet another team coming from the nearest accessible railway station at Nenana, 674 miles away. The plan was for the two packs to meet roughly in the middle at Nulato and hand over 300,000 units of life-saving serum.

When you look at the map, it is a mammoth distance and then you have to consider the conditions. Gale-force winds and blizzards meant that vision was reduced to almost nothing.

In charge of one of the teams of dogs was leading dog-breeder and musher, Leonhard Seppala. He had moved from his native country of Norway to Nome during the gold rush and his lead sleigh dog was Togo, named after Japanese Admiral Togo Heihachiro.

Togo was small (weighing only forty-eight pounds) but deceptively strong. He had been a sickly puppy and his disobedient temperament did not seem suited to his becoming a working dog, not least when he kept getting into various scrapes with bigger dogs. He followed Seppala at every opportunity and when his owner tried to give him away as a pet, Togo jumped out of a window to get back to him. Eventually, Seppala bowed to the inevitable and put him in a harness. The transformation in his temperament was a revelation. Togo settled down immediately. It was as if he had been making mischief just to get himself noticed, and now he had found his true calling. Seppala described him as 'an infant prodigy' and a 'natural born leader'.

Togo and Leonhard Seppala who called him 'the best dog that ever travelled the Alaska trail'.

Togo was already twelve years old by the time he took part in the gruelling five-and-a-half-day mission to get the serum back to Nome. As the temperature dropped below –60°F, several of the mushers lost fingers or toes to frostbite. One man needed hot water poured over his hands to release them after they froze to the sleigh handle. Several of the 150 dogs died of exhaustion and cold.

All across America, people anxiously followed the team's progress in what was called the 'Great Race of Mercy'. The longest and hardest stretch was led by Togo and Seppala. Together they crossed the frozen seas of the Norton Sound during a blizzard in the dead of night. They climbed the 5,000 feet of the Little McKinley mountain and travelled more than 260 miles in three days to carry the serum to the next team in the relay.

Gunnar Kaasen and his dog Balto then led the final fifty-five-mile leg of the journey, arriving in Nome in the early morning of 2 February. They had almost lost their precious cargo at the eleventh hour, when their sleigh overturned and Kaasen got frostbite as he searched through thick snow to find the metal canister containing the vials of medicine.

Balto, who brought the serum into Nome, was feted for saving the population. He was the one who got all the newspaper attention and a statue in New York's Central Park. Togo, who had run more than four times further in far more treacherous conditions, was forgotten. Seppala was heartbroken, saying:

> I hope I shall never be the man to take away credit from any dog or driver who participated in that run. We all did our best. But when the country was roused to enthusiasm over the serum run driver, I resented the statue to Balto, for if any dog deserved special mention it was Togo.

No one dog should get all the credit. It took a whole team of them to pull off the remarkable and life-saving delivery.

Togo did finally get some public acclaim, making appearances to large crowds across the USA. He was also awarded a gold medal for bravery – fittingly presented by the man who had given his master his first sleigh team, Roald Amundsen himself. To Seppala it was too little too late: 'I never had a better dog than Togo. His stamina, loyalty and intelligence could not be improved upon. Togo was the best dog that ever travelled the Alaska trail.'

Togo died in December 1929 at the age of sixteen and remained forever in Seppala's thoughts. About thirty years later, when he was eighty-one, Seppala wrote in his journal: 'When I come to the end of the trail, I feel that along with my many friends, Togo will be waiting and I know that everything will be all right.'

TOMMY

In 1942, a blue cock pigeon became an accidental spy. His name was Tommy (or NUPR.41.DZ56) and his story begins with a race that went wrong.

While many birds were volunteered for the war effort and signed up to the National Pigeon Service, others were held on to by their owners for the duration. One of these was Tommy, who was a racing pigeon of repute. But when his breeder, William Brocklebank from Dalton-in-Furness, Cumbria, entered him into a contest in Cheshire, things went awry: Tommy was blown off course on his way home from Nantwich and ended up near Santpoort, in the Nazi-occupied Netherlands.

New laws meant that the keeping of pigeons was illegal in the Netherlands and the authorities were killing all racing pigeons to prevent anyone from using them to carry secret messages out of the country. Luckily for Tommy, he was found by a postman who had friends in the Dutch Resistance, a group of civilians working under cover to make life as hard as possible for the troops who had invaded their country. The postman passed the exhausted bird on to one of them, pigeon enthusiast Dick Drijver, who did his best to nurse Tommy back to health.

Drijver had kept birds of his own until the Germans occupied Holland in 1941 and ordered all homing pigeons to be killed and their identifying leg rings handed in to the authorities. He did everything he could for Tommy, assuming that he was an English bird who would know how to find his way back home, though he was not sure the pigeon would ever fly again.

But when he received an urgent message that needed to reach British forces as quickly as possible, he released Tommy with it, hoping that this time he would find his way home safely. The note he had attached to the bird's leg detailed critical information about a Nazi munitions factory near Amsterdam.

It was a perilous 400-mile journey. Tommy was shot in the wing by a German marksman but managed to continue flying, making it back to Dalton against all the odds and arriving with blood dripping from his breastbone. Brocklebank promptly passed the note to police and, two days later, a broadcast on BBC Dutch Radio Service News let the Resistance know that the critical mission had been a success – all thanks to one rather battered and bruised pigeon.

Drijver continued his Resistance work until he was arrested

by the Gestapo and sent to a concentration camp. Somehow he managed to jump from the train and remained in hiding for the rest of the war. Tommy made a full recovery and became quite the celebrity, with people paying to meet him at agricultural shows around the country.

William Brocklebank (left) with Tommy, on the day he was awarded the Dickin Medal and Dick Drijver was presented with a pair of English pigeons.

The two were finally reunited when Tommy was awarded the Dickin Medal in 1946 'for delivering a valuable message from Holland to Lancashire under difficult conditions, while serving with NPS in July 1942'. They were presented with a medal by the head of the Dutch secret service and Dick was given a pair of English pigeons to take back to the Netherlands to re-establish his own loft.

TRIM

Melbourne is one of my favourite cities in the world. I first went there to cover the Melbourne Cup for BBC Radio in 1999 and I made sure I went back regularly until the Commonwealth Games in 2006. I was part of the TV team for the BBC and was assigned the night shift, which was great in terms of audience (everyone at home would be awake) but hideous in terms of lifestyle (I was working while everyone was asleep). I devised a plan to keep me mentally alert and happy. I would sleep as late as I could and then go for a walk from the Broadcast Centre down to Flinders Street Station, pick up a healthy smoothie and walk back to start writing scripts and catch up on all the action. I did not then realise that the beautiful art nouveau building, which dates back to 1909, was named after Matthew Flinders, the man who mapped Australia. Nor did I realise that he might not have been able to do it without the invaluable assistance of a cat called Trim.

Captain Matthew Flinders from Lincolnshire was a navigator and cartographer who, in 1801, led the second circumnavigation of what was then known as 'New Holland'. He was the one who identified it as a continent for the first time, calling it Terra Australis which became Australia.

It was common in those times to have a cat on board. Not only would they keep down any rodent infestation, but they were thought to bring good luck and good weather. They were also great company for the crew. Hence two-year-old Trim set sail alongside Flinders on the HMS *Investigator*. After falling overboard on a previous voyage, he had impressed everyone by swimming back and climbing a rope to safety. Flinders thought he would make the perfect companion for his most ambitious voyage to date and named him accordingly after Corporal Trim, an eccentric servant and sidekick in Laurence Sterne's novel *Tristram Shandy*, which was a huge bestseller at the time.

Trim was the admiral of the ship's cats. He enjoyed an elevated position at the captain's right hand, dining alongside him at his table. He liked to keep himself busy, so the crew taught him some great tricks. He learnt to leap over the deckhands and to play dead, lying on his back with his paws stretched out until a signal was given for him to get up again. He enjoyed playing with a musket ball strung up with twine and at one point took a fancy to nautical astronomy, observing the measurements being taken with 'earnest attention'.

Trim involved himself in every aspect of life on board, often from a lofty perch looking down as though he were in charge. Flinders wrote:

> This assumption of authority to which, it must be confessed, his rank, though great as a quadruped, did not entitle him amongst men, created no jealousy; for he always found some good friend to caress him after the business was done, and to take him down in his arms.

Their exploration and mapping eventually complete, the pair set off to return to England on HMS *Porpoise* but they ran aground on the Great Barrier Reef and they were all, including Trim, forced to swim to safety on a small island where they remained stranded for seven weeks before rescue arrived. Trim remained stoical at all times and did his best to keep their spirits up.

The voyage home was eventually resumed, but the timing wasn't good. England and France were at war and when they stopped at the Isle de France (Mauritius) for supplies and essential repairs, Flinders was arrested on suspicion of spying.

He was kept under house arrest for the next six years but put his time to good use, recording the details of his exploration and writing odes to Trim. While Flinders stayed imprisoned, there was no holding Trim back. He would head off to explore Mauritius until one night in 1804 he did not come back. Flinders was convinced he had been captured and eaten. He wrote:

> Thus perished my faithful intelligent Trim! The sporting, affectionate and useful companion of my voyages during four years. Never, my Trim, 'to take thee all in all, shall I see thy like again', but never wilt thou cease to be regretted by all who had the pleasure of knowing thee.

Flinders vowed that if he made it home safely, he would erect a monument to 'perpetuate thy memory and record thy uncommon merits'.

Ill health meant he never got the chance to do so, but wherever he is celebrated today, Trim is too. There's a lovely statue of Flinders in Donington, Lincolnshire, the town of his birth, and snuggling up against his leg is a cat. Similarly in Sydney, outside the Mitchell Library, there is a statue of Flinders and behind it on the window ledge there is a bronze of an alert, agile and adventurous cat called Trim.

UGGIE

From double rejection to the Hollywood Walk of Fame sounds like the précis for a film pitch and in a way, it is. Uggie may be an icon rather than a hero but he represents the many animals who have starred on screen and not received the credit they deserve.

Show-business dogs have long been stars in Hollywood. Most famous of all is probably Pal, the Collie who played Lassie in the original film. The most successful canine actress of her time was Terry the Cairn Terrier who made sixteen film appearances, the most high-profile being Toto in *The Wizard of Oz*. At $125 a week, she earned more on *The Wizard of Oz* set than most of the human actors. Judy Garland, who played Dorothy, became so attached to Toto that she wanted to adopt her but Carl Spitz, who owned and trained her, refused.

There was also the German Shepherd Rin Tin Tin, a world-famous silent-movie star, who, it was rumoured, received more votes in the Best Actor category at the Oscars than any of his human rivals. But it was deemed 'unseemly' for a dog to win and, instead, the award was given to the German actor Emil Jannings for his roles in *The Last Command* and *The Way of All Flesh*.

More recently it has been Uggie, a Parson Russell Terrier, who rose from humble beginnings to become a cause célèbre for those who believe that great performance should be rewarded, regardless of whether the actor has two legs or four.

Uggie had a difficult start in life, commenting in his autobiography:

> I think I met my father once when he came to sniff dispassionately at me and my sprawling siblings. All that I remember of my mother was that she was gentle and nurturing; the smell of warm milk would forever remind me of her. Sadly, I was plucked from her teat early on and sold to the first stranger to pick me out from the litter.

As a puppy, Uggie was so wild and boisterous that he was rejected by two sets of owners and seemed destined for the dog pound. Hollywood dog trainer, Omar von Muller, heard about the unmanageable mutt, and took him in until a new home could be found. The dog was a handful but von Muller spotted something in him that made him distinctive: 'He was a crazy, very energetic puppy, and who knows what would have happened to him if he [had] gone to the dog pound. But he was very smart and very willing to work.' Von Muller also noticed that he had the precious characteristic of not being afraid of things. He explained:

> That is what makes or breaks a dog in the movies, whether they are afraid of lights, and noises and being on sets. He gets rewards, like sausages, to encourage him to perform, but that is only a part of it. He works hard.

Uggie was one of seven dogs living in North Hollywood with von Muller and his family. The others worked in the film industry and so it was no surprise when Uggie followed that career path. Like most actors, he worked his way up by doing bit parts and commercials. It wasn't until he was nine, in 2011, that he hit the big time. First, he got the role of Queenie in *Water for Elephants*, alongside Reese Witherspoon and Robert Pattinson. This was followed by a show-stealing performance in the silent film *The Artist* with Jean Dujardin.

Uggie stars in The Artist *in 2011.*

Uggie played the part to perfection, doing most of his own stunts, despite the fact he had two doubles. At the American Film Institute premiere, he walked the red carpet with the rest of the cast and was called upon to promote the film just as much as either of his two-legged co-stars. There were photo shoots and television appearances galore, including a trip to London to appear on the Graham Norton show and to attend a charity screening in aid of the Dogs Trust.

The critics were unanimous in their praise: 'A dog whose IQ seems to be higher than that of most actors of any species', reported the *Daily Telegraph*; 'A scene-stealing terrier', said *Rolling Stone*; 'Keep your eye on the dog!' added CNN; while the *New York Post* film critic Lou Lumenick said that Uggie had turned in 'the best performance, human or animal, in any film I've seen this year'.

Uggie's rave reviews turned him into a superstar and would have surely earned any human actor a plethora of award nominations. Those who believed he was being ignored launched a campaign called 'Consider Uggie'. S. T. VanAirsdale, an editor at *Movieline*, who believed that Uggie's performance had outshone Leonardo DiCaprio's in *J. Edgar*, took to Facebook to drum up support with the backing of *The Artist*'s cast and crew. The Academy was having none of it and although the film went on to win five awards, including Best Picture, Best Actor and Best Director, Uggie's name was left firmly off the list.

The *Daily Telegraph* took up Uggie's cause, saying that a win for him would be a win for all the dogs who had received plaudits for their big-screen performances, but it was still not to be. When BAFTA members asked whether they could vote for Uggie as Best Actor for the 2012 awards, they were told, 'Regretfully, we must advise that as he is not a human being and, as his unique motivation as an actor was sausages, Uggie is not qualified to compete for the BAFTA in this category.' I don't understand why sausages should be deemed unacceptable motivation when money works for humans.

Uggie did receive acclaim when he won the Palm Dog Award at the Cannes Film Festival in 2011, 'for one of the best performances ever in the history of the award', as well as praise at the Golden Collar Awards. His paw prints were memorialised in cement outside Grauman's Chinese Theatre in LA, keeping company with other impressions from the great and the good, including one of Whoopi Goldberg's dreadlocks, Groucho Marx's cigar, Betty Grable's legs and Marilyn Monroe's hands. It was the first-ever canine paw-print ceremony and doubled as Uggie's official retirement-from-the-movies party, with a cake in the shape of a fire hydrant.

By this stage, Uggie was hot property but he was also almost ten years old – seventy in dog years – and von Muller felt that he should start taking it easy and walk away from the fifteen-hour days on set. It was a no to any more big parts and, instead, Uggie became the spokesdog for both Nintendo (where he promoted the video game Nintendogs and Cats) and People for the Ethical Treatment of Animals (PETA), where he helped to encourage people to adopt dogs from shelters.

Uggie also found time to write his autobiography, with the assistance of Wendy Holden who said, 'I thought this was the one Hollywood star I really wanted to write about.' While she admitted that he 'channelled his thoughts through Omar', she said, tellingly, 'He was ready to talk.'

His fans were also ready to listen and very keen to meet him. Uggie and Holden went on a book tour, where

> people were literally queuing around the block to see this tiny furry star. There was something about him that changed people. Women especially adored him. People approached him far more readily than a human star.

Uggie's book is dedicated to his *Water for Elephants* co-star Reese Witherspoon, with the words: 'For Reese, my love, my light'.

When Uggie died in 2015, Reese Witherspoon paid tribute on Twitter, calling him a 'special, sweet soul'. Von Muller called him 'a perfect little terrier. I will forever hold him dearly in my heart and never forget his infinite love for chicken and hot dogs.' PETA issued the statement:

> His remarkable life is a reminder that countless dogs and cats are waiting in animal shelters for someone to 'discover' them. Like Uggie, each of them is a star – they just need a loving home where they can shine.

Regrettably, Uggie was excluded from the obituary section at the Oscars that year. The Academy once again failed to recognise his artistic contribution, even in death.

Other canine movie stars who must be mentioned include Bingo, the Otterhound, who starred as Sandy in the 1982 original film of *Annie*. Butkus Stallone, real-life pet of Sylvester, inspired the *Rocky* screenplay and appears in the film as the boxer's training partner. Beethoven the St Bernard stole our hearts in the film of that name, while the 1974 film *Benji*, about a stray who just wants to have a home, still makes me cry. Hooch gave the Dogue de Bordeaux newfound levels of fame with his comedy turn alongside Tom Hanks in *Turner and Hooch*.

Closer to home, a Belgian Tervuren found fame as Wellard in *EastEnders*; Shep, Petra and Goldie all enjoyed long stints as *Blue Peter* stalwarts; and Buster the Boxer crashed the internet when the John Lewis Christmas advert for 2016 showed him bouncing on a trampoline.

UNSINKABLE SAM

They say that every cat has nine lives, but perhaps none has put that to the test quite as seriously as Unsinkable Sam. This was a cat who survived not one, not two, but an unprecedented three shipwrecks during the Second World War.

Like Tirpitz, Sam started his military career with the Kriegsmarine, or the German navy, before fate intervened and he swapped sides to join the Allies. He set sail on board the *Bismarck* as it left to take part in Operation Rheinübung. It didn't go well. Nine days later, on 27 May 1941, the battleship was sunk amidst fierce fighting. Of 2,100 crew, only 115 survived. One was a black-and-white cat, found floating on a board and picked up by HMS *Cossack*.

In the International Code of Signals, the letter 'O' stands for 'man overboard', so the crew named him Oscar. Or occasionally 'Oskar', as he was German by birth. Over the next few months, Oscar enjoyed life as the ship's mascot as the *Cossack* carried out convoy duties in the Mediterranean and North Atlantic. Their peace was shattered on 24 October 1941 when a torpedo from German submarine U-563 badly damaged the *Cossack* as she escorted a convoy from Gibraltar back to Britain. A third of the forward section was blown off, killing 159 men, and leaving the ship listing badly. Plans to tow it back to Gibralter were scuppered by poor weather. The crew was speedily transferred to HMS *Legion*, and on 27 October the *Cossack* sank. After surviving this second shipwreck, the crew renamed Oscar Unsinkable Sam.

Sam's next home was the aircraft carrier HMS *Ark Royal*. That didn't last long either. On 14 November 1941 another German U-boat hit the target with another torpedo. Again, attempts to save the ship proved futile. She sank thirty miles off shore. Sam seemed to be making a habit of floating on planks of wood and waiting to be saved. Along with the other survivors, he was picked up by a motor launch, 'angry but quite unharmed'.

It's hard to say whether Sam was lucky, unlucky or simply

a jinx on all who sailed with him, but his spirit of survival made him exceptional. He happily gave up his ocean-going adventures to become chief mouser at the offices of the governor of Gibraltar before living out the rest of his life at a home for sailors in Belfast.

Other cats, too, made their names by surviving shipwrecks. My favourite story is that of ship's cat Maizie who helped the crew of her vessel keep calm and carry on.

Maizie's merchant ship was sunk in the North Atlantic in March 1943. Together with six crew members, she spent fifty-six hours on a life raft before being rescued and taken to safety. Several of the men suffered from exposure or seasickness, and Maizie took time to comfort each in turn, 'almost like a mother'.

Along with the men who lived, she survived by eating malted milk biscuits and other meagre rations. One of them later said, 'If Maizie hadn't been with us we might have gone nuts. We completely forgot our personal discomfort and almost fought for the privilege of petting her.'

VALEGRO

A year before the London Olympics, I was asked which sport would be the 'dark horse' of the Games. I immediately (and with a complete lack of irony) answered: 'Dressage'. You see, I had seen Valegro in action and I knew that even non-horsey people would be captivated by his elegance, his rhythm, his fluency and the remarkable partnership between him and his rider Charlotte Dujardin.

The art and discipline of dressage goes back to Ancient Greece, where Xenophon of Athens wrote a treatise on horsemanship in 355 BC. He realised that in any cavalry battle, the horses who could swerve sideways, turn on a sixpence, change direction or leap out of the way would be at a huge advantage. Dressage was the foundation of equine warfare.

During the Renaissance, these ideas were rekindled and it became very much the trend for European monarchs to be painted on horseback looking as if they were ready to do dressage. For the Holy Roman Emperor Charles V or Philip II of Spain, Catherine the Great of Russia or Charles I of England and Scotland, it was deemed an essential skill to be able to ride well; portraits on horseback added to the monarchy's elevated status. In France, Louis Henri, Duke of Bourbon, Prince of Condé took the notion even further by becoming convinced he would be reincarnated as a horse and therefore built himself the largest and most ornate palace for horses. The stables at Chantilly were completed in 1740. They remain in all their glory as a museum to the horse and are the backdrop to the racecourse that stages both the French Derby and the French Oaks.

At around the same time in Vienna, Emperor Charles VI commissioned a large indoor hall to be converted into a riding school. It was opened in 1735 and became the new home of the Spanish Riding School (which had already existed for 200 years and specialised in training a breed of horse from Spain called the Lipizzaner). The Spanish Riding School still gives performances today with the highlight being the 'Airs Above the Ground', a routine which has the dancing horses leap into the air.

In the twentieth century, dressage developed into a sport and was included in the Olympics in Stockholm in 1912, although only military officers were allowed to compete until after the Helsinki Olympics in 1952. The sport has been variously compared to horse ballet, dancing or gymnastics and is an advanced and highly skilled form of riding that tests both horse and rider. Together, they are 'expected to perform from memory a series of predetermined movements' with absolute precision. Moves include the piaffe, when the horse trots on the spot, its hooves almost hanging in the air; the pirouette, when it does a tight circle in canter, its hind legs staying in the same place; and the amazing flying changes, when it changes lead leg every stride in canter. Top-class Grand Prix dressage is a beautiful thing to behold and when you add music and original choreography for the Freestyle routines, it's like watching a show in the West End.

In London 2012, the star of stage and screen was Valegro. Born in 2002 in the Netherlands, he was one of several horses that came over to join the stable owned by regular British team rider Carl Hester. He was bought for just £4,000 – top dressage horses can cost up to £100,000 – and nicknamed Blueberry. The edible nickname turned out to be highly appropriate for a horse who has always loved his food. When it looked as though he wasn't going to be tall enough for Hester to ride himself, he was almost sold on again, twice. As luck would have it, a new groom had just started at Hester's yard in Gloucestershire and she was the perfect size for Blueberry. So it was that Charlotte Dujardin found herself riding the horse of a lifetime.

Their partnership was based on hard work and mutual trust and it was clear from a very early stage that they were made for each other. They worked their way through every level at the National Championships and although the plan was for Hester to take over when they got to Grand-Prix level, he could see that this was too good a relationship to tear apart.

Valegro was ten and Dujardin twenty-seven years old when they made their Olympic debut in London. They set a new Olympic record score in the Grand Prix with 83.74 per cent and led the team (which included Hester) to a first gold medal for Great Britain in dressage. Two days later Dujardin and Valegro performed a crowd-pleasing patriotic Freestyle routine to music that included 'Land of Hope and Glory' and the chimes of Big Ben. Watching the performance live in Greenwich made people weep and it still gives me goosebumps today. It was the perfect combination of power, precision and passion, all woven together with fluency and flair. The judges were impressed and the pair galloped to a convincing individual gold medal with a score of 90.089 per cent.

Between 2012 and 2016, Valegro raised the bar again and again, winning more World and European titles and setting world-record scores in Grand Prix, Grand Prix Special and Freestyle. At the Rio Olympics, the pressure was on to defend their Olympic title. The heat and the long journey did not work in Valegro's favour, but he and Dujardin secured the individual gold medal and helped the team to a silver medal.

Valegro was retired in December 2016, after a faultless Freestyle performance at the Olympia London International Horse Show. He had won three Olympic gold medals, two world titles, seven European gold medals and two dressage World Cups. Dujardin wept as she paid tribute to him:

> I've had an incredible time. He's given me the best journey and I have achieved more than I thought I

> would ever achieve with him. I cannot thank him enough. He's the horse everyone dreams of having, the perfect horse. He is just magical and he has captured so many people's hearts.

That's where Valegro stands out: he made people who didn't understand – or had never watched – dressage care about it, enjoy it and appreciate it. Hester also paid tribute to his extraordinary career:

> I can't really explain what he has done for not only British dressage, but the world of dressage. He is a phenomenon, and to have a horse like this in your life is very special. But to us at home he is still very much a family pet; he doesn't know he is worth as many millions as people say. He is just a very kind, loving horse, who likes nothing more than to eat grass and looks forward to mealtimes.

Valegro is still enjoying his food in retirement and is ridden out every day to keep him fit and happy.

VOGELSTAR

One of the wonders of the avian world is created by starlings: the murmuration, when as many as 100,000 birds come together to form expanding, contracting and ever-changing aerial formations. It is thought to be a defence against attacks from predators such as birds of prey. I remember standing at

RSPB Minsmere watching one. I've never seen anything like it. The way the starlings move in waves as if choreographed to perform: it's almost as if they are dancing to music, each one reacting in a split second to the movements of the birds closest to them. In parts of Denmark more than a million can be seen flying together in what is known as *sort sol* or 'black sun'. I was mesmerised at this wonder of the natural world, which got me thinking about starlings.

Starlings are not popular with farmers or airline pilots. A large flock can cause a plane to crash if they get into the engines – a so-called 'bird strike'. They will also swoop onto farmland and eat crops or feed meant for livestock. They are estimated to cause up to $800 million in damage to the US farming industry per year.

On the bright side, they are highly intelligent and very good mimics. They can copy a tune, another bird's voice, the croak of a frog, a car alarm or a phone's ringtone. It is this skill that brought a starling to the attention of one of the world's greatest composers.

In April 1784 in Vienna, Mozart had just put the finishing touches to his Piano Concerto No. 17 in G. He was planning the inaugural public performance in June. Meanwhile, out shopping, he was stunned to hear a caged starling singing the motif from the concerto's Allegretto (the third movement). It was almost note perfect. He bought the bird on the spot and named it Vogelstar (the German word for starling).

Mozart's paranoia that his work might be copied by a fellow composer was legendary, and some historians report that he was immediately worried about how the bird could possibly know parts of a piece that was yet to be unveiled. Others assume that Mozart had taught the starling the piece while he was in the shop, or that the piece had been performed earlier than is recorded.

He noted the price he paid for the bird (thirty-four kreuzer) in a ledger, along with the transcriptions of both versions of the tune and the words '*das war schön*' ('that was wonderful').

Mozart and his pet starling lived happily together for three years. As the birds are noisy at best and known for singing along to music whenever they can, it could have been a distraction but, instead, they clearly formed a deep bond.

When his overbearing father died in 1787, the composer did not travel the 250 miles to his funeral in Salzburg. Some cite their difficult relationship as reason for this, although Mozart wrote to a friend at the time, 'You can imagine the state I am in.' When the starling died just weeks later, Mozart held a formal ceremony for the bird in his garden, instructing attendees to wear velvet capes in its honour. Notes from his friend Georg von Nissen record, 'He arranged a funeral procession, in which everyone who could sing had to join in, heavily veiled.' Mozart erected a proper gravestone in the garden of his lodgings on which he had written an inscription. He had also written an elegy in tribute to the bird:

Here rests a bird called Starling,
A foolish little Darling.
He was still in his prime
When he ran out of time,
And my sweet little friend
Came to a bitter end,
Creating a terrible smart
Deep in my heart.
Gentle Reader! Shed a tear,
For he was dear,

Sometimes a bit too jolly
And, at times, quite folly,
But nevermore
A bore.

After his double bereavement, Mozart's next completed work was A *Musical Joke*, K522. This piece is seen as a parody of the work of rival composers but I wonder if he wrote it as a way of cheering himself up at a time of grief. I am very familiar with the final movement of the piece because it's the theme music for the BBC's equestrian coverage. I will now hear it as a tribute to Mozart's beloved starling.

WARRIOR

Warrior by name and warrior by nature, this was 'the horse the Germans couldn't kill'. He survived bombs, bullets and burning buildings. Together with Jack Seely, who bred and owned him, Warrior spent four years on the front lines and survived the whole of the First World War.

Warrior was a bay, thoroughbred gelding born in the Isle of Wight in 1908. He was strong, powerful and fast with a fine, handsome head. He was not big at 15.2 hands, but he was nimble and had character. When Seely first sat on him, the two-year-old bucked him off three times in a row, but after that the bond they forged was unbreakable.

Warrior and Seely were part of the original British Expeditionary Force, arriving in France on 11 August 1914. Seely had served in the Boer War and while in South Africa was elected as an MP: first, as a Conservative in 1900, and later as a Liberal. He was a great friend of Winston Churchill and was promoted to Secretary of State for War just before the fighting broke out.

The First World War was the last war in which horses played a major part. Their role changed as the conflict progressed, as weaponry development left them increasingly vulnerable on the battlefield. The German Army only used horses on the Western Front at the start of the conflict, as did the United States, though British forces deployed mounted troops from the beginning of the war to the end. At the start, they had only 25,000 horses, so the Horse Mobilisation Scheme was introduced and over 100,000 animals were compulsorily purchased to boost their numbers. Over the course of the war, between 500 and 1,000 horses were shipped out to Europe every day. Britain lost one horse for every two men, a total of 484,000.

Horses offered a number of advantages to the rudimentary mechanised vehicles of the time. They could be used to pull ambulances or artillery in difficult settings over rough terrain or through thick mud. They also helped to boost morale among the troops. Horses were used for cavalry charges and for ferrying supplies, for reconnaissance and for carrying messages or wounded soldiers. But they paid a heavy price for their versatility. They were no more immune to disease or exhaustion, to the dangers of poison gas or artillery fire, than the men fighting alongside them, and hundreds of thousands lost their lives.

Like so many other horses, Warrior experienced the worst of war. At Passchendaele the mud was almost his undoing. The ground around him was littered with dead horses, who had been unable to free themselves from the mire. Warrior, too, sank up to his belly. It took four men to dig him out and was, in Seely's words, 'a narrow escape'. The pair faced countless enemy charges, were buried under debris and subjected to shelling, but Seely said:

> never once did he attempt to bolt or to do any of the things which might be expected of an animal reputed to be so naturally timid as the horse. No, my stout-hearted horse not only kept his own fear under control but by his example helped beyond measure his rider and his friend to do the same.

When an explosion set Warrior's stable on fire, trapping him under the burning beams, he managed to escape. When a shell landed on the ruined cottage that was being used as his temporary home, he made it out from under the rubble. The fierce fighting on the Somme and at Ypres claimed more than 150,000 British lives, with hundreds of thousands more wounded and missing, and yet somehow Warrior survived. When a German shell broke in two, close to where he was tethered, one part hit a horse nearby and cut it clean in half. Warrior was unharmed.

The only time Seely rode another horse – Warrior was lame – the animal was hit by a shell and killed instantly. Seely suffered three broken ribs and said a silent thank you that he had not been on his beloved Warrior that day. On another occasion, a German sniper shot at Warrior but narrowly missed, instead killing the horse who had been nose to nose with him. Warrior was like a cat with nine lives.

As an officer, Seely usually rode at the head of the column

so he and Warrior were first in the line of fire. The troops fighting alongside him continued to be inspired by his bravery. He was their good-luck charm and they patted his flank whenever they passed him.

The last major cavalry charge of the war came at Moreuil Wood in March 1918. Seely (by now General Seely) and Warrior led the men of the Canadian Cavalry Brigade into a battle where a quarter of the men and half of the horses did not survive to tell the tale. The pair followed the leading tank, when an explosion saw it crash into the canal as the bridge they were crossing collapsed. They were unharmed and nothing could deter the horse from charging into the thick of the fighting. Seely remembered that the horse

> was determined to go forward and with a great leap started off. All sensation of fear had vanished from him as he galloped on at racing speed. There was a hail of bullets from the enemy as we crossed the intervening space and mounted the hill, but Warrior cared for nothing.

The fighting got ever more brutal and the casualty list continued to grow, as whole battalions fell. At Gentelles, Seely was gassed and both his replacement horses were killed but he and Warrior made it through relatively unscathed.

The pair returned home at Christmas 1918. Warrior would enjoy a long and happy life at his home on the Isle of Wight. Four years later, he even won a local point-to-point. When he died at the age of thirty-two, it was reported in *The Times* and the *Evening Standard*. His portrait was painted many times, famously by Sir Alfred Munnings, and a small bronze statue of him and Seely stands near his home. A century after he went off to the front, Warrior was posthumously awarded an honorary Dickin Medal on behalf of all the animals who served in the First World War.

WHEELY WILLY

I am horrified at the cruelty humans will inflict on animals and I'm afraid this story of a heroic animal is also a story of man's inhumanity. How could anyone cut the throat of a tiny Chihuahau to sever his vocal chords? How could they abuse him so badly that he became paralysed from behind the front legs to the tip of his tail? What on earth is going through the mind of someone who does this to a helpless little dog?

This dog, who became known as Wheely Willy, survived for his tale to be told. He is a symbol of courage and kindness, and the message he embodied struck a chord with children across America and around the world. The message was this: 'Life is what you make of it.'

Willy was left for dead in a cardboard box on a street in Los Angeles. He was found by a passer-by and taken to a local animal hospital, but it didn't look good. He couldn't bark or whine because of the brutal removal of his vocal chords. Vets were eventually able to stabilise his condition with surgery and other treatment, but it was clear that he would never walk again. His fur had been shaved and he was so emaciated that he was constantly cold and shivering. So much so that the staff nicknamed him Chilly Willy.

Despite the trauma and the cruelty he had endured, the tiny dog proved to have an indomitable spirit. It took him a year to recover from his injuries and he then faced another challenge. No one wanted to take on a pet with such profound disabilities. It looked as though there was no other option than for him to be put to sleep.

Deborah Turner, who ran a dog-grooming parlour, heard of his plight and wanted to help. She went to visit him, planning to tell his story, but when she saw how cheery he was – determined to play despite dragging his legs behind him – she decided to take him home with her. She adopted Willy and vowed to do all she could to make his days active and happy.

The little dog had a real zest for life, despite everything he had been through. He soon made friends with Turner's other pets, which included dogs, cats and a turtle, but he couldn't keep up with them. Turner realised she needed to find a way to make him mobile again. Initial experiments using a skateboard and helium balloons failed. The plan had been to raise his back legs off the ground and onto the board, but Willy was so tiny the balloons lifted his whole body off the ground!

This required something a little more sophisticated. K9 carts were developed in the USA in the 1960s for a range of animals (from rats to miniature horses) to help restore mobility if they had been injured. Turner ordered a cart for Willy. It was a two-wheeled brace that would support his hindquarters and allow him to use his front legs to propel himself around. Well, Willy loved it and took off apace. 'It took his world from black and white into full-blown colour', Turner said.

When his story featured in the local paper, Turner put

Willy's new-found fame to good use by taking him to see Alzheimer's patients and those on psychiatric wards. Turner described him as 'enabled' rather than 'disabled'. The little dog on wheels empowered everyone who met him or read about him, including veterans who had suffered spinal-chord injuries. Willy still couldn't speak – his bark sounded more like a frog than a dog – but *Animal Planet* deemed him 'Wheely Willy', the motivational speaker. He made various TV appearances and started a world tour, going round schools and hospitals. Deborah could see the incredible impact he had on children with injuries or disabilities: 'They could see that having a disability was not the end. By looking at Willy they realised you could be disabled and still have a happy life', she explains.

Wheely Willy was particularly popular in Japan: his fans included Prince Hitachi and Princess Hanako, who got down onto the floor to greet him, causing something of a sensation.

Willy died in 2009, but will always be remembered for offering hope to so many people. He proved that it was possible to go through devastating injury or illness and still come out the other side to find a new way of living life to the full.

WOJTEK

As a little girl, I was not into dolls. I did, however, love teddy bears. Warm, furry, cuddly and comforting, they were so much more fun than hard plastic figures with long blonde hair. I had an impressive collection of bears of various types – panda bears, polar bears, black bears in particular, fat little Winnie the Pooh bears. I loved Winnie the Pooh and took to collecting stationery, T-shirts, pictures and, of course, all of the books. My brother was more of a Paddington fan. He had many Paddington bears and he took to wearing a duffle coat and carrying a small leather briefcase. I think in his imagination, he *was* Paddington.

From a boy who thought he was a bear to a bear who thought he was human. Wojtek was the bear who won the Second World War. Seriously.

The story of Wojtek begins at a small railway station on the Iran/Iraq border in 1942. A group of Polish soldiers got out of a truck to stretch their legs. They were on a long journey south from the gulags of Siberia, where they had been deported when Stalin had invaded East Poland in 1939. Now that Russia was at war with Germany, they had been released and were returning to join the British Army in Egypt to fight the Nazis.

They were in the mood for celebration and entertainment, so when a young shepherd boy approached them with an orphaned bear cub in a sack, it felt like a good omen. The deal required some cash, some chocolate and a Swiss army knife, and suddenly they had a new mascot and pet. The bear cub was skinny and underfed, so the soldiers shared their rations with him. He particularly liked condensed milk, which they fed him from an old vodka bottle.

After three months in a Polish refugee camp near Tehran, the growing cub was donated to the 2nd Transport Company. It was unprecedented to have a bear at the centre of a major tactical and operational unit of the Polish armed forces, and the soldiers named their new comrade Wojtek.

Found when a cub, Wojtek grew to over six feet tall and was extremely comfortable in human company.

The bear took to army life with aplomb. He learnt to follow Polish instructions and to greet his seniors by saluting them. He was not always perfectly behaved and developed a habit of stealing the soldiers' kit from the clothes lines to play with and swing round his head. He also caused havoc in the food store, devouring the planned Christmas Eve feast in 1942.

On one occasion, however, his misdeeds earned him only plaudits. In the scorching heat of the Middle East, the bear loved nothing more than to take a refreshing shower. He had mastered getting into the shower huts and would whine until a sympathetic comrade would allow him a turn of the nozzle. Eventually, he learnt how to turn on the flow by himself so the doors were kept locked to deter him.

One night in June 1943, he spotted one of the doors to the huts ajar and he lost no time in lumbering inside. Hiding in the shower was a spy who had sneaked into the camp to steal ammunition. He was not expecting 440 pounds of bear to pop in for a wash and, paralysed with fear, he screamed the place down – making it easy for Wojtek's fellow soldiers to catch him. The bear was rewarded with cigarettes, which he mainly ate, and beer, which became his favourite tipple. He dined on fruit, syrup and (like Paddington) handfuls of marmalade. There was no argument about his taking a long, refreshing shower that night.

Polish II Corps was posted to North Africa for final training and then, in 1944, sent to Italy to fight alongside the British Eighth Army. There was no question that their bear would be going too, but there was a strict rule prohibiting pets getting close to any of the action. The troops came up with an ingenious solution: if the bear joined up, he would be eligible for the same rights as the rest of them. Wojtek thus became an official serving soldier of the Polish II Corps (later the 22nd Artillery Supply Company). He was treated in the same way as the rest of his comrades. He had the rank of corporal and a serial number. He got paid, was given double rations and slept in a tent. The only difference was that in between missions he was put in a large wooden crate rather than crammed into the back of a truck with his fellow soldiers. His arrival from Egypt was duly effected although it did cause some confusion when the officer processing the Polish soldiers called Wojtek's name, and there was no response.

The Poles formed an important part of the Allied campaign in Italy and their fighting spirit was much admired. In the words of an Irish Guards officer in the 78th Division: 'Of their resolve there was no doubt. For whose gallantry the Division soon learnt to feel an awed yet amused admiration. They exposed themselves with the most reckless abandon. They seem to know no fear.' The same could be said of their bear, who by now was six feet tall and weighed thirty-five stone.

Many years later, British Courier Archibald Brown said in an interview, 'We looked at the roster, and there was only one person, Corporal Wojtek, who had not appeared.' When he asked the other soldiers why he hadn't come forward, one replied, 'Well, he only understands Polish and Persian', and then led the officer to the cage holding the bear.

Wojtek could do the work of four men. At the bloody Battle of Monte Cassino, he helped move 100 pounds of artillery shells to an advanced position. He seemed unperturbed by the tremendous danger and the death toll: 55,000 Allied soldiers were killed before the Germans were defeated. It was the Poles who finally captured the ancient abbey of Monte Cassino, by then bombed to bits, where they raised their flag. Wojtek's strength and bravery had proved so vital to the cause that a new emblem was designed for the 22nd Company of a bear carrying an artillery shell.

The Yalta Agreement in 1945 saw the Soviets impose Communist rule over Poland, and Stalin's regime left many Poles unable or unwilling to return to their homeland. Having continued to fight alongside the Allies, many eventually chose to settle in Britain. The Transport Corps were among them, Wojtek included, and they made their home in Berwickshire.

Wojtek led a victory parade down Princes Street in Edinburgh before he settled into a life at Winfield Airfield near Hutton in the Scottish Borders. He earned his keep, carrying heavy boxes and logs wherever needed and, in his downtime, grew partial to a swim in the River Tweed.

When the Polish Army was finally disbanded in 1947, Wojtek was found a new home at Edinburgh Zoo, where his keepers indulged the bear's taste for milky tea. He was visited often by his former comrades who would throw him treats of cigarettes and chocolate for old time's sake, and he always perked up when he was addressed in Polish.

Wojtek died in 1963, at the age of twenty-three. His extraordinary contribution to the Allied war effort was marked with a plaque at the Imperial War Museum as well as a statue in Edinburgh's West Princes Street Gardens, another one in Duns in the Borders and one in Krakow, Poland.

All four tributes are a vivid testament to how much the British owed the people of Poland during the Second World War, and just how much one friendly bear changed the lives of his soldiers. In the words of one of his former comrades, Ludwik Jaszczur, 'I'll tell you the truth. Wojtek helped us to win the Second World War.'

ZARAFA

What do you get as a present for the king who has everything? A giraffe, of course. Which is why, in 1827, Ottoman ruler Muhammad Ali of Egypt sent giraffes to three European monarchs as diplomatic gifts. This caused something of a stir, not least because a giraffe had not been seen in Europe since 1486, and never in France.

Here's a random fact: did you know that giraffes are the champions of power naps? They have to be alert to predators all the time so they can't afford to lie down and fall into a deep sleep. I can fall into a really deep slumber in the back of a car, on a train or on a plane within seconds. It's always been a useful way of getting through a long day when I need to be on form and full of energy, but I'm not sure I'd be able to survive without my solid eight or nine hours every night. If I was a giraffe, I'd be easy pickings.

Giraffes are, of course, known not only for their height, but also their distinctive spotted coats. The scientific name for the giraffe – *Giraffa camelopardalis* – references the fact that many believed the beast to be a cross between a camel and a leopard. While their coats might look similar, no two giraffes have exactly the same pattern.

Zarafa – the giraffe earmarked for King Charles X of France – had an epic journey all the way from Sennar, Sudan, where she had been captured by Arab hunters, to Paris. It involved camels, a boat down the Blue Nile to Khartoum, another barge all the way to Alexandria and a voyage across the Mediterranean Sea to Marseille. Nineteenth-century ships were not built for such tall cargo, so a hole was cut on deck for Zarafa to poke her head through and a canvas tent erected to shade her from the elements.

Three cows also on board provided plenty of liquid refreshment for the crew and their special cargo and, after a month at sea, she became the first giraffe to set foot on French soil, on 31 October 1826. She spent the winter in Marseille, while people puzzled over the easiest and safest way to get her to the capital. The young giraffe was good natured and friendly, making her easy to handle, but she was also frisky and energetic. Her enthusiasm often led her to break into a trot, or even a gallop and, as giraffes can reach speeds of thirty-eight miles per hour, this made it a bit tricky for her handlers, who often got dragged along with her. It was finally decided that the best solution to get her to Paris was to walk the 550 or so miles to her new home.

Naturalist Etienne Geoffroy Saint-Hilaire, director of the zoo at the Jardin des Plantes, Paris, decided to join her, ensuring she was kitted out for the trek with specially made shoes. In case of inclement weather, she had a black raincoat emblazoned with the fleur-de-lis, King Charles's coat of arms.

At the age of fifty-five, and suffering from gout and rheumatism, Saint-Hilaire made an unlikely companion, but nonetheless, when the warmer weather arrived, in May 1827, they set off on their epic expedition north. The procession, complete with cows, an antelope and a police escort, was quite a spectacle.

Unlike most other quadrupeds, giraffes move both legs on one side at the same time, giving them a rather lurching or wobbly-looking gait. But this certainly didn't hamper the young calf's progress: she kept up a steady two miles an hour throughout, and seemed to thrive. 'She gained weight and much more strength from the exercise: her muscles were more defined, her coat smoother and glossier', reported Saint-Hilaire.

Their journey took them through the towns and villages of France, drawing a crowd of enthusiastic fans wherever they went. When she reached Lyon on 6 June, a 30,000-strong crowd – almost a third of the city's population – turned out to greet her. The cavalry had to be brought in to help the mounted police keep order.

She was dubbed '*la belle Africaine*' by the press, while Saint-Hilaire referred to her as '*le bel animal du roi*'. She would later become known as 'Zarafa', a variant of the Arabic word for giraffe (*zerafa*) meaning 'charming' or 'lovely one'.

It took forty-one days to reach Paris, where the giraffe was presented to the king at the Château of Saint-Cloud on 9 July 1827, nibbling rose petals from his hand before settling in to her new home in the Jardin des Plantes.

An exotic thirteen-foot-tall, long-necked, spotted animal from Sudan was a must-see for any self-respecting Parisian. In the last three weeks of July alone, she was visited by more than 60,000 people, among them the authors Honoré de Balzac (who wrote a story about her) and a very young Gustave Flaubert. The magazine *La Pandore* reported that 'the giraffe occupies all the public's attention; one talks of nothing else in the circles of the capital'.

Zarafa's distinctive coat was considered the epitome of chic and spawned a fashion craze – '*la mode à la girafe*' – which swept the country. Women wore their hair in towering topknots to simulate her horns, even though this looked ridiculous, and made their hair so tall that they could only travel by carriage if they sat on the floor. There were accessories galore, including hats, combs and ties, and the season's 'in' colour was a shade of yellow known as 'belly of giraffe'. Spotted fabrics were sported by the great and the good, and giraffe-themed wallpaper ended up on the walls of many a high-class salon. Giraffes featured in pretty much everything from topiary to gingerbread.

The fashion craze came to an end around the same time as the reign of Charles X (the July Revolution of 1830 saw him abdicate and flee to Britain), but Zarafa stayed in Paris. She lived out the rest of her days peacefully, away from the jostling crowds and gawpers that had been such a feature of her youth. She remained a mythical, unforgettable figure. One commentator wrote, 'one is seized by astonishment at the sight of her', and so it was that she was remembered for leaving the people of France with a new sense of awe at the wonders of the animal kingdom.

ACKNOWLEDGEMENTS

It's been an absolute joy to write this book and I've learnt a huge amount along the way. It's been a pleasure to work with a team as professional and as committed as those at John Murray, and I thank them for keeping me occupied during what would otherwise have been a rather empty year.

A book that contains this much information needs a whole team to make it possible, and I owe huge thanks to Cari Rosen for her detailed research, Georgina Laycock for the concept and her brilliant editing, Candida Brazil for her careful copy-editing, Howard Davies for his eagle-eyed proof-reading, Juliet Brightmore for sourcing the fabulous photos and Caroline Westmore for her attention to detail.

They say you should never judge a book by its cover, but if you picked this one up because you liked the jacket, I will not condemn you. Instead, I will pay due credit to Sara Marafini who designed it, Kate Brunt who organised and directed the photo shoot and Paul Stuart who always gets the best shots – this time with the willing co-operation of Sherlock the fire dog, who was better than any model. Many thanks to his handler, Station Commander Paul Osborne, for allowing Sherlock an afternoon in front of the camera.

Thanks to Janette Revill for her creativity in design and Diana Talyanina for producing the book so beautifully, as well as Ruth Ellis for her indexing ingenuity. Rosie Gailer and Jess Kim will make sure it gets all the publicity and marketing it needs, and Ellie Wheeldon has had her ears on the audio version.

My literary agent Eugenie Furniss has impeccable judgement and when she said this was a good idea, I didn't hesitate to commit. She and her assistant Emily MacDonald have been hugely supportive.

Writing a book requires focus and dedication. Neither of these comes naturally to me, so I am indebted to Alice for giving me the required kick up the backside when needed, or appearing with a cup of coffee to check I hadn't been distracted by shopping for random kitten toys on the internet.

Finally, I'd like to thank you for reading this, and I hope in doing so you have appreciated anew how much animals have always enriched our world and how much they are part of our cultural history. We human beings have not always deserved their bravery and devotion, but I hope that we will all do more to protect, preserve and honour their lives.

PICTURE CREDITS

Dedication page: © Paul Stuart. Introduction: Author's collection. *Aldaniti*: Mirrorpix/Alamy Stock Photo. *Alex*: Gerald Davis/Shutterstock. *Antis*: Courtesy of Nina Buckley and Magdalena Luscombe. *Aramis and Brothers*: Koen Van Weel/EPA/Shutterstock. *Arkle*: Alamy Stock Photo. *Barrie*: @sean_lilo and @barriespaws. *Barry (St Bernard)*: Chronicle/Alamy Stock Photo. *Bob*: © Clint Images/Hodder & Stoughton Publishers. *Bobbie the Wonder Dog*: newspaper cutting of 1924. *Bucephalus*: © The Mullan Collection/Mary Evans Picture Library. *Canaries*: Steve Bent/Associated Newspapers/Shutterstock. *Charisma*: *NZ Horse & Pony Magazine*, Manaia Media Ltd, New Zealand. *Cher Ami*: Imago History Collection/Alamy Stock Photo. *Chips*: Bettmann/Getty Images. *Christian*: John Rendall/AAP/PA Images. *Clara*: The Picture Art Collection/Alamy Stock Photo. *Congo*: World History Archive/Alamy Stock Photo. *Daisy*: Medical Detection Dogs. *Diesel*: AP/Shutterstock. *Dolly*: Jeremy Sutton Hibbert/Shutterstock. *Félicette*: © CNES/CERMA, 1963. *Finn*: Shutterstock. *Gallipoli Murphy*: Chronicle/Alamy Stock Photo. *Gander*: Library and Archives Canada/Department of National Defence fonds/a116791. Photo: Item Number 1180. *Geese*: The Picture Art Collection/Alamy Stock Photo. *G. I. Joe*: TopFoto/PA Images. *Greyfriars Bobby*: Chronicle/Alamy Stock Photo. *Gustav*: AP/Shutterstock. *Hachiko*: CPA Media Pte Ltd/Alamy Stock Photo. *Ham*: Science History

Images/Alamy Stock Photo. *Hoover*: Bill Ryerson/The Boston Globe/Getty Images. *Huberta*: Front cover of *The Saga of Huberta* by G. W. R. Le Mare, 1931. *Jack (dog)*: Pictorial Press Ltd/Alamy Stock Photo. *Jack (baboon)*: Wikimedia Commons/Public Domain. *Jake*: Alan Diaz/AP/Shutterstock. *Jet*: AFP/Getty Images. *Jovi*: Courtesy of Graham Sage. *Judy*: TopFoto. *Kika*: © Pete Summers/SWNS. *Koko*: Keystone Pictures USA/Shutterstock. *Laika*: Detlev Van Ravenswaay/ Science Photo Library. *Learned Pig*: Mary Evans Picture Library/Harry Price. *Lin Wang*: Historic Collection/Alamy Stock Photo. *Lonesome George*: Mark Green/Alamy Stock Photo. *Magic*: Gentle Carousel Miniature Therapy Horses, www.gentlecarouseltherapyhorses.com. *Mick the Miller*: Topical Press Agency/Getty Images. *Milton*: Split Seconds/ Alamy Stock Photo. *Minnie*: Courtesy of Major Maurice Taylor XXth LF. *Miracle Mike*: Bob Landry/The LIFE Picture Collection/Getty Images. *Moko*: Gisborne Herald/AP/ Shutterstock. *Molly*: Colin Butcher UKPD, United Kingdom Pet Detectives. *Mr Magoo*: *Duluth News Tribune*, MN, USA. *Paddy the Wanderer*: Alexander Turnbull Library, Wellington, NZ (Ref ½-122301-F). *Paul*: Patrik Stollarz/AFP/Getty Images. *Pickles*: DPA/PA Images. *Pudsey*: Shutterstock. *Rats*: Chamussy/Shutterstock. *Reckless*: PDSA. *Red Rum*: Keystone Press/Alamy Stock Photo. *Rifleman Khan*: PDSA. *Rip*: Albatross/Alamy Stock Photo. *Rocky*: Invicta Kent Media/ Shutterstock. *Sadie*: PA Images. *Salty and Roselle*: Richard Drew/AP/Shutterstock. *Secretariat*: Bettmann/Getty Images. *Sefton*: Popperfoto/Getty Images. *Sergeant Bill*: Public Domain. *Sevastopol Tom*: © The National Army Museum/Mary Evans Picture Library. *Sherlock*: © Paul Stuart. *Shrek*: Shutterstock. *Simon*: PDSA. *Siwash*: Wallace Kirkland/The LIFE Picture Collection/Getty Images. *Smoky*: Albatross/Alamy Stock

Photo. *Stoffel*: Moholoholo Wildlife Rehabilitation Centre, Limpopo Province, South Africa, www.moholoholo.co.za. *Stubby*: Science History Images/Alamy Stock Photo. *Tacoma*: Training dolphins in the Persian Gulf/US Navy/Public Domain. *Tama*: Toru Yamanaka/AFP/Getty Images. *Tamworth Two*: Mike Forster/Daily Mail/Shutterstock. *Thandi*: Kariega Game Reserve, Eastern Cape, South Africa, www.kariega.co.za. *Thula*: www.irisgracepainting.com. *Tirpitz*: Chronicle/Alamy Stock Photo. *Togo*: George Rinhart/Corbis via Getty Images. *Tommy*: Len Cassingham/Associated Newspapers/Shutterstock. *Trim*: Frank Augstein/AP/Shutterstock. *Uggie*: AA Film Archive/Alamy Stock Photo. *Valegro*: David Rogers/Getty Images. *Vogelstar, Starlings*: Tom Langlands Photography/Alamy Stock Photo. *Warrior*: Canadian War Museum/Beaverbrook Collection of War Art/CWM19710261-0450. *Wheely Willy*: Koichi Kamoshida/Getty Images. *Wojtek*: © Imperial War Museum/HU 16546. *Zarafa*: Chronicle/Alamy Stock Photo. *'la mode à la girafe'*: *The Repository of Arts, Literature, Fashions, &c.*, vol. XII, July 1828, R. Ackermann, London. *Animal silhouettes*: Shutterstock.com.

INDEX

See Contents, pages vii–x, for the 100 heroic animals.